PALGRAVE STUDIES IN THE HISTORY OF
SCIENCE AND TECHNOLOGY

Series Editors

James Rodger Fleming (Colby College) and Roger D. Launius (National Air and Space Museum)

This series presents original, high-quality, and accessible works at the cutting edge of scholarship within the history of science and technology. Books in the series aim to disseminate new knowledge and new perspectives about the history of science and technology, enhance and extend education, foster public understanding, and enrich cultural life. Collectively, these books will break down conventional lines of demarcation by incorporating historical perspectives into issues of current and ongoing concern, offering international and global perspectives on a variety of issues, and bridging the gap between historians and practicing scientists. In this way they advance scholarly conversation within and across traditional disciplines but also to help define new areas of intellectual endeavor.

Published by Palgrave Macmillan:

Continental Defense in the Eisenhower Era: Nuclear Antiaircraft Arms and the Cold War
By Christopher J. Bright

Confronting the Climate: British Airs and the Making of Environmental Medicine
By Vladimir Jankovic´

Globalizing Polar Science: Reconsidering the International Polar and Geophysical Years
Edited by Roger D. Launius, James Rodger Fleming, and David H. DeVorkin

Eugenics and the Nature-Nurture Debate in the Twentieth Century
By Aaron Gillette

John F. Kennedy and the Race to the Moon
By John M. Logsdon

A Vision of Modern Science: John Tyndall and the Role of the Scientist in Victorian Culture
By Ursula DeYoung

Searching for Sasquatch: Crackpots, Eggheads, and Cryptozoology
By Brian Regal

Inventing the American Astronaut
By Matthew H. Hersch

The Nuclear Age in Popular Media: A Transnational History
Edited by Dick van Lente

Exploring the Solar System: The History and Science of Planetary Exploration
Edited by Roger D. Launius

The Sociable Sciences: Darwin and His Contemporaries in Chile
By Patience A. Schell

The First Atomic Age: Scientists, Radiations, and the American Public, 1895–1945
By Matthew Lavine

NASA in the World: Fifty Years of International Collaboration in Space
By John Krige, Angelina Long Callahan, and Ashok Maharaj

Empire and Science in the Making: Dutch Colonial Scholarship in Comparative Global Perspective
Edited by Peter Boomgaard

Anglo-American Connections in Japanese Chemistry: The Lab as Contact Zone
By Yoshiyuki Kikuchi

Eismitte in the Scientific Imagination: Knowledge and Politics at the Center of Greenland
By Janet Martin-Nielsen

Eismitte in the Scientific Imagination

Knowledge and Politics at the Center of Greenland

Janet Martin-Nielsen

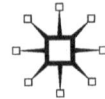

EISMITTE IN THE SCIENTIFIC IMAGINATION
Copyright © Janet Martin-Nielsen, 2013.

All rights reserved.

First published in 2013 by
PALGRAVE MACMILLAN®
in the United States—a division of St. Martin's Press LLC,
175 Fifth Avenue, New York, NY 10010.

Where this book is distributed in the UK, Europe and the rest of the world, this is by Palgrave Macmillan, a division of Macmillan Publishers Limited, registered in England, company number 785998, of Houndmills, Basingstoke, Hampshire RG21 6XS.

Palgrave Macmillan is the global academic imprint of the above companies and has companies and representatives throughout the world.

Palgrave® and Macmillan® are registered trademarks in the United States, the United Kingdom, Europe and other countries.

ISBN: 978–1–137–38079–1

Library of Congress Cataloging-in-Publication Data

Martin-Nielsen, Janet, 1982–
    Eismitte in the scientific imagination : knowledge and politics at the center of Greenland / Janet Martin-Nielsen.
        pages cm.—(Palgrave studies in the history of science and technology)
    Includes bibliographical references.
    ISBN 978–1–137–38079–1 (alkaline paper)
    1. Scientific expeditions—Greenland—History. 2. Research—Greenland—History. 3. Ice sheets—Greenland—History. 4. Greenland—History. 5. Greenland—Geography. 6. Greenland—Discovery and exploration. 7. Scientific expeditions—Political aspects—Denmark—History. 8. Scientific expeditions—Political aspects—France—History. 9. Scientific expeditions—Political aspects—United States-History. 10. Geopolitics—History. I. Title.

G760.M325 2013
919.8204—dc23   2013025586

A catalogue record of the book is available from the British Library.

This book is printed on paper suitable for recycling and made from fully managed and sustained forest sources. Logging, pulping and manufacturing processes are expected to conform to the environmental regulations of the country of origin.

Design by Newgen Knowledge Works (P) Ltd., Chennai, India.

First edition: December 2013

10 9 8 7 6 5 4 3 2 1

For my little boy, Lars Ole

Contents

List of Maps	ix
Acknowledgments	xi
Major Archival Collections Consulted	xiii
List of Abbreviations	xv
Introduction The Edge of the World, the End of the World	1
Chapter 1 A Land Apart	11
Chapter 2 Taming the Ice Sheet	39
Chapter 3 The Longest Trek	61
Chapter 4 It Has Completely Changed	85
Epilogue A Conspicuous Absence	115
Notes	123
Bibliography	181
Index	203

Maps

1 Early journeys on Greenland's ice sheet, including the Wegener expedition's 1930–1931 trek to Eismitte — xvii
2 Expéditions Polaires Françaises in Greenland, 1948–1953 — xviii
3 Project Jello's 1955 trek and major US facilities in Greenland — xix
4 Expédition Glaciologique Internationale au Groënland, 1956–1960 — xx
5 The HIRAN stations on Greenland's ice sheet, established in 1956 — xxi

Acknowledgments

This book is the product of three wonderful years of postdoctoral work at Aarhus University's *Center for Videnskabsstudier* (Centre for Science Studies). It is a pleasure to thank my advisor, Matthias Heymann, for untiring advice, good humor, and a hearty dose of patience. Ronald E. Doel's enthusiasm and expertise, too, have been invaluable, and back home in Canada my doctoral advisors—Brendan Gillon, Mark Solovey, and Janis Langins—have continued to provide warm mentorship.

This book and the ideas which underlie it have benefited from the advice and encouragement of many colleagues. From discussions at conferences (particularly in Aarhus, Copenhagen, Stockholm, and Oslo) to detailed readings of the manuscript, the book is stronger for their efforts. I thank Dania Achermann, Mark Carey, Ronald E. Doel, Matthew Farish, Kristine C. Harper, Matthias Heymann, Christian Kehrt, D.J. Kinney, Henrik Knudsen, John Krige, Trevor Levere, Henry Nielsen, Kristian Hvidtfelt Nielsen, Christopher Jacob Ries, Peder Roberts, and Sverker Sörlin. The reviewers of this manuscript and of my related papers helped refine my thinking and set my narrative in its larger context. Mark Solovey's ongoing guidance about the publishing process has also been a welcome support.

I have been blessed throughout the research and writing of this book with archivists who have been not only knowledgeable, but also both kind and genuinely interested in my work, repeatedly going out of their way to help me access materials. I thank them immensely: in Denmark, the Arktisk Institut in Copenhagen (especially Vibeke Sloth Jakobsen), the Rigsarkivet in Copenhagen, and the Erhvervsarkivet in Aarhus; in France, the Archives Nationales de France in Fontainebleau; in Belgium, the NATO Archives in Brussels (Victor Martinez-Garzón); in Canada, Library and Archives Canada in Ottawa (Kathy Chow); and in the United States, the US Army Corps of Engineers' CRREL Library in Hanover, New Hampshire (Elizabeth Hoffmeister), the National Archives and Records Administration in College Park, Maryland, the

Roger G. Barry Archives at the National Snow and Ice Data Center in Boulder, Colorado, the American Geosciences Institute in Alexandria, Virginia (Karin Mills), the Mandeville Special Collections Library at the University of California at San Diego (Heather Smedberg), the Byrd Polar Research Center Archival Program at Ohio State University (Laura Kissel), the National Oceanic and Atmospheric Administration Central Library in Silver Spring, Maryland (Albert "Skip" Theberge), and the National Technical Information Service in Alexandria, Virginia.

At Aarhus University, Susanne Nørskov has the uncanny ability to find the most obscure of sources based on minimal information—half a title here, an author's misspelled surname there—and make them appear on my desk, always accompanied by the widest of smiles. Special thanks to Jim Kinter for keeping alive the history of US Air Force photomappers and surveyors, and to Roy P. Velvick for generously sharing memories of his time in Greenland. And to the colleagues who offered materials and assisted with the retrieval of archival documents, especially when I passed the best before date for flying overseas while pregnant: Ronald E. Doel, Kristine C. Harper, Christian Kehrt, Maiken Lolck and Henry Nielsen, your assistance is much appreciated.

The Carlsberg Foundation, via the *Exploring Greenland: Science and Technology in Cold War Settings* project, provided the funding which made the research and writing of this book possible. At Palgrave Macmillan, Chris Chappell and Sarah Whalen have been paragons of efficiency throughout the publication process, and have responded to a plethora of questions from me with infinite patience. And at Aarhus University, Minna Elo and Lene Bongaarts made living in a fifth country in nine years as seamless as could be: *tusen tak*. This book was spoken with Dragon Dictate voice recognition software.

At the end of the day, it is family and friends upon whom we rely. My parents, Ole and Kathy Nielsen, and my brother, Steve, made me realize early on that academia was for me. Many years ago, they lived in the same Aarhus apartment in which much of this book was written. Margaret Curtis's late night forecasting shifts and early morning balloon launches stateside mean that I always have a sounding board, regardless of time zones. My husband, Richard, believed entirely in this project from the very beginning, often with more conviction than I could muster. It is thanks to him that the technical aspects of the book are in order – and, more importantly, that I was so well taken care of during the writing process. And my son, Lars Ole, born just after I finished writing chapter 3: he has been with me the whole way through.

JANET MARTIN-NIELSEN
June 2013

Major Archival Collections Consulted

(followed by abbreviations used in references, where applicable)

- Archives Nationales de France (Fontainebleau, France)
  - Expéditions Polaires Françaises Collection, 1914–2001
- Arktisk Institut—Polarbiblioteket (Copenhagen, Denmark)
- Library and Archives Canada (Ottawa, Canada) (LAC)
  - Denmark and Greenland General File
  - Arctic Institute of North America Collection
- National Snow and Ice Data Center—Roger G. Barry Archives (University of Colorado at Boulder, Boulder, CO, USA) (NSIDC)
  - Carl S. Benson Collection
- North Atlantic Treaty Organization Archives (Brussels, Belgium) (NATO)
- Rigsarkivet (Danish National Archives, Copenhagen, Denmark)[1]
  - Grønlandsministeriet (Greenland Ministry) Collection
  - Udenrigsministeriet (Ministry of Foreign Affairs) Collection
- United States National Archives and Records Administration (College Park, MD, USA) (NARA)
  - General Records of the Department of State
  - Records of the US Joint Chiefs of Staff
  - Records of Headquarters US Air Force
  - Records of the Army Staff
  - Records of the Weather Bureau (Geophysical Year)
- US Army Corps of Engineers—Cold Regions Research and Engineering Laboratory Library and Archives (Hanover, NH, USA) (CRREL)

Abbreviations

ADTIC	Arctic, Desert and Tropic Information Center
AINA	Arctic Institute of North America
APCS	Air Photographic and Charting Service
CRREL	Cold Regions Research and Engineering Laboratory
DEW	Distant Early Warning
EGIG	Expédition Glaciologique Internationale au Groënland (International Glaciological Expedition to Greenland)
EPF	Expéditions Polaires Françaises (French Polar Expeditions)
GISP	Greenland Ice Sheet Project
HIRAN	HIgh-precision shoRAN
IGY	International Geophysical Year
LAC	Library and Archives Canada
NARA	National Archives and Records Administration
NATO	North Atlantic Treaty Organization
NORAD	North American Aerospace Defense Command
NSIDC	National Snow and Ice Data Center
SHAPE	Supreme Headquarters Allied Powers Europe
SHORAN	SHOrt-RAnge Navigation
SIPRE	Snow, Ice and Permafrost Research Establishment
WGS	World Geodetic System

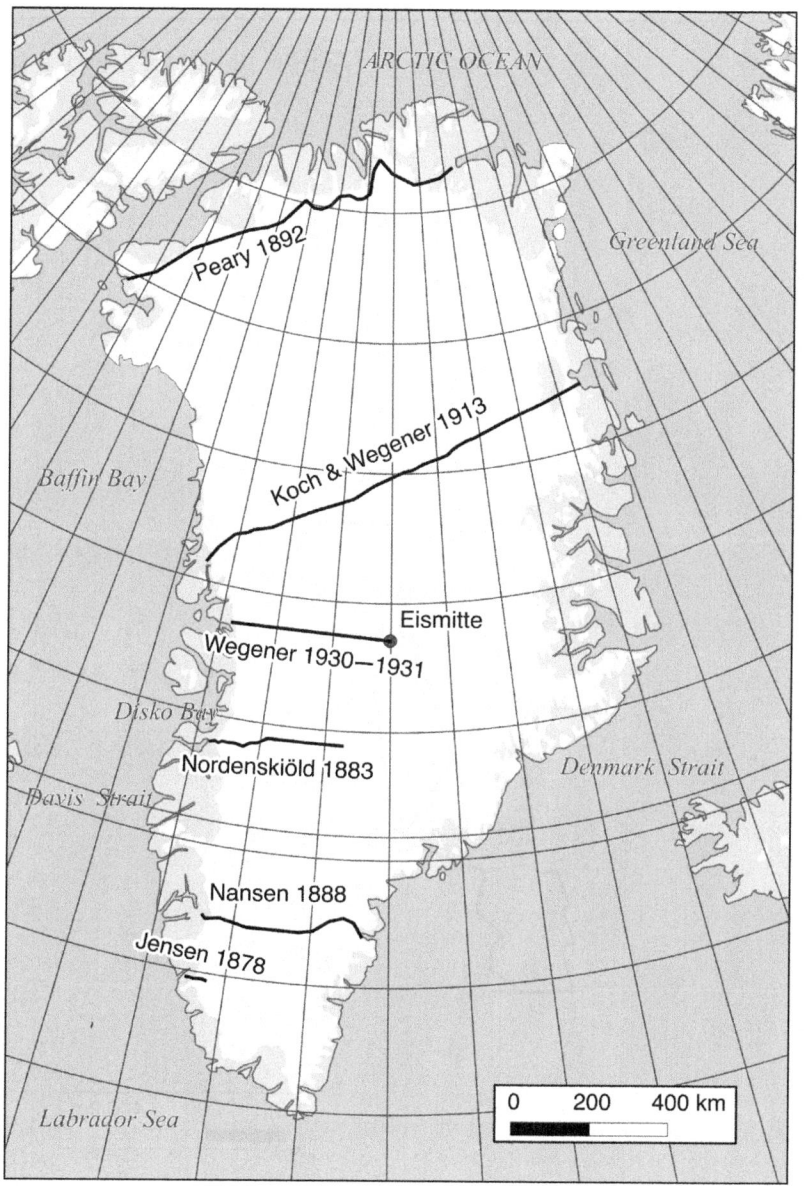

Map 1 Early journeys on Greenland's ice sheet, including the Wegener expedition's 1930–1931 trek to Eismitte.

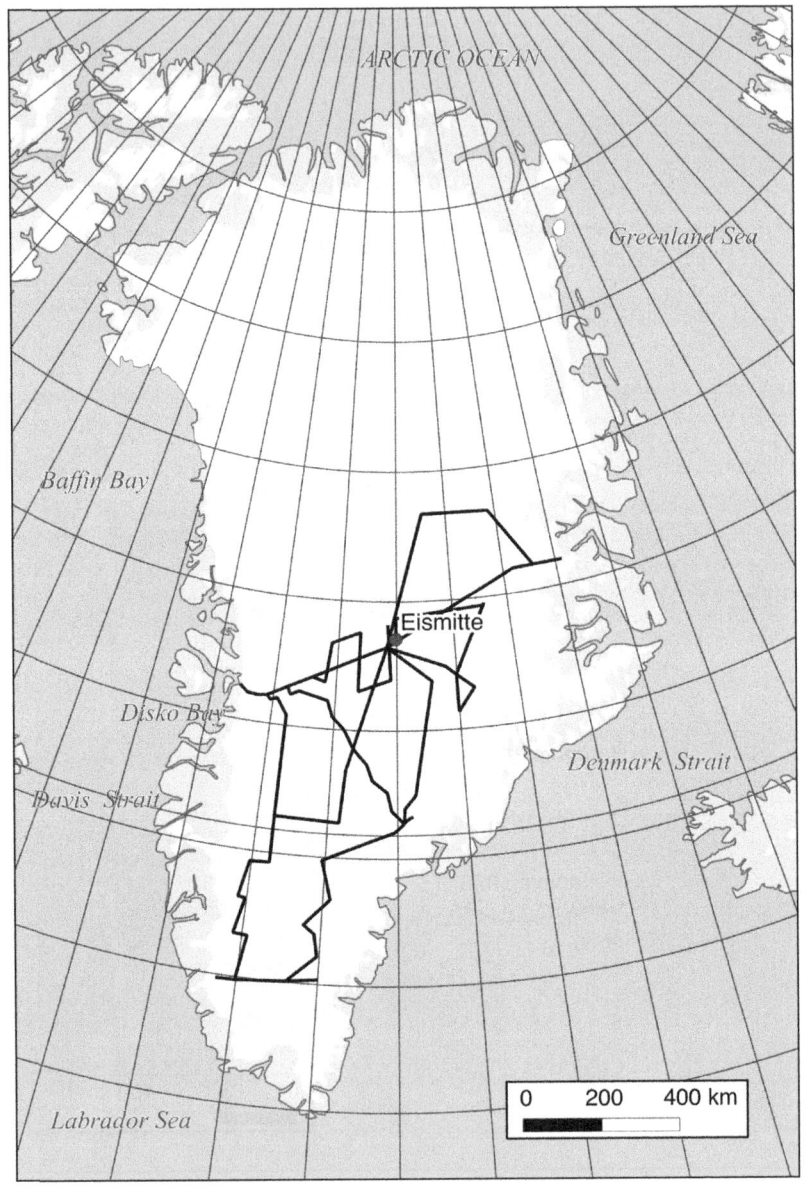

Map 2 Expéditions Polaires Françaises in Greenland, 1948–1953.

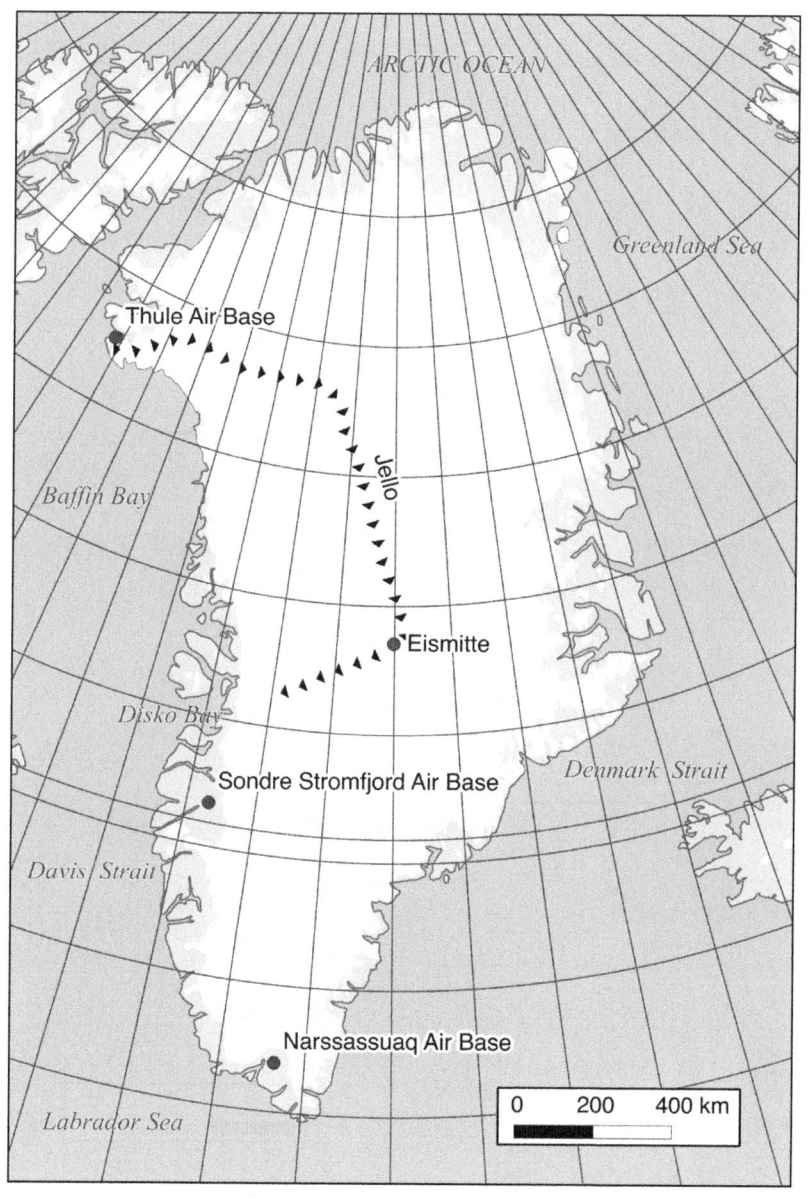

Map 3 Project Jello's 1955 trek and major US facilities in Greenland.

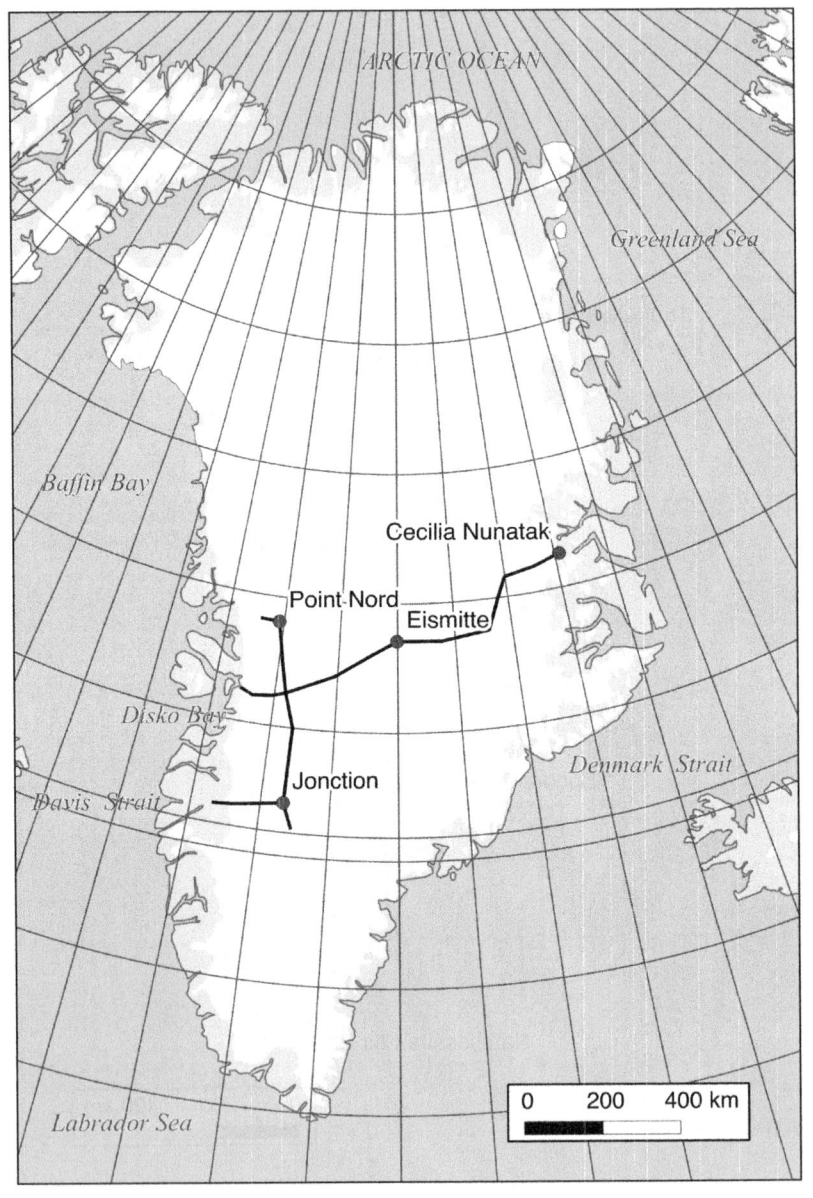

Map 4 Expédition Glaciologique Internationale au Groënland, 1956–1960.

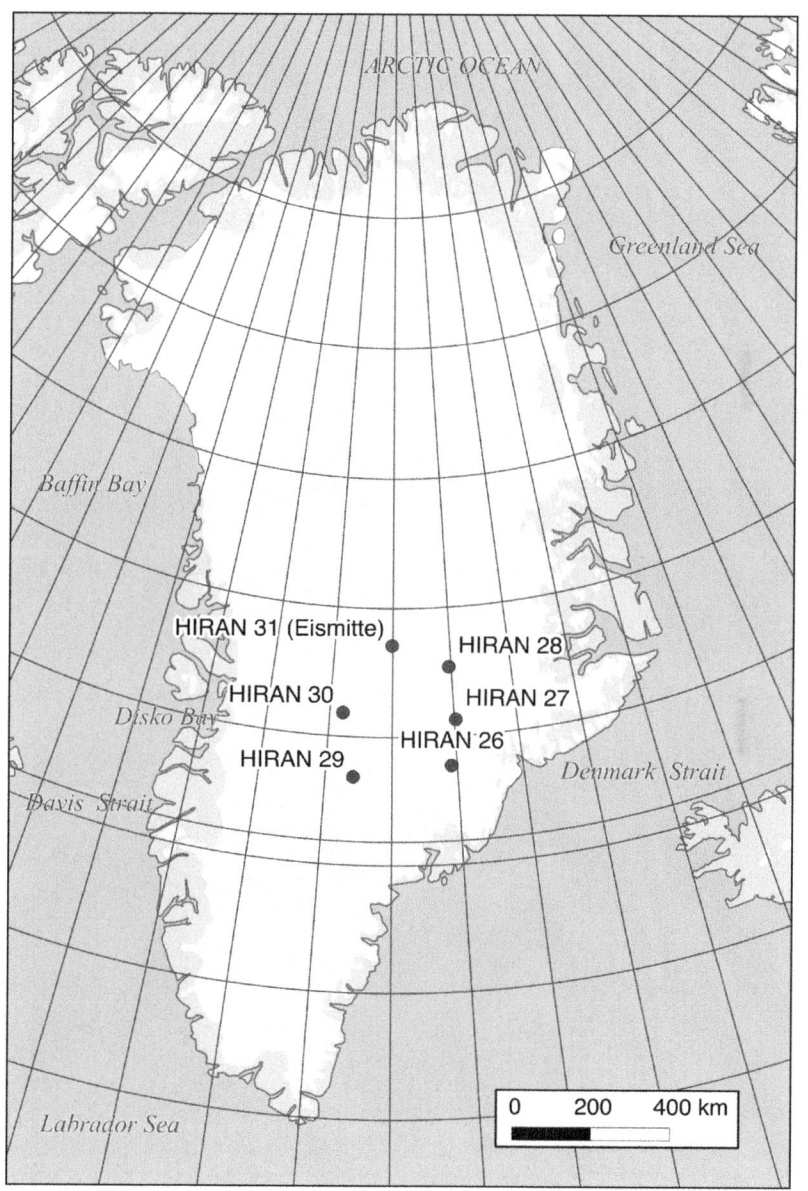

Map 5 The HIRAN stations on Greenland's ice sheet, established in 1956.
All maps created by Richard Martin-Nielsen using Natural Earth and Quantum GIS.

Introduction

The Edge of the World, the End of the World*

71° 8' North, 40° 3' West: Eismitte, literally *middle ice*, the geometric center of Greenland's ice sheet. *"When we leave our tent and turn towards the east, it is 500 kilometers of uniform ice desert separating us from the Greenland Sea. When we turn towards the west, it is 500 identical kilometers to the Baffin Sea from which we came. Towards the south? 1,200 kilometers, similarly uniform. Towards the north? 1,200 kilometers, without an undulation. [It is] the edge of the world, the end of the world. But not even, because this is not the world that man knows..."*[1] Remote and inhospitable as it is, Eismitte represents one of the scientifically richest locations in Greenland, and indeed in the Arctic world. This book traces the scientific history of Eismitte. It begins with the first efforts to penetrate into Greenland's interior in the eighteenth century and ends with the postwar race to map the world fueled by the development of intercontinental ballistic missiles. In between, it examines the first overwinter at the center of the ice sheet, spearheaded by famed German polar researcher Alfred Wegener in 1930–1931, the *Expéditions Polaires Françaises* (EPF) (French Polar Expeditions) expedition of 1948–1953, the United States military 1955 Project Jello trek, and the cooperative European *Expédition Glaciologique Internationale au Groënland* (EGIG) (International Glaciological Expedition to Greenland) expedition of 1956–1960.

What drew so many men, from so many countries and over so many years, to such a desolate location? Eismitte's attraction, of course, was not singular. For Wegener and his German colleagues, the center of the ice sheet represented at once a great polar challenge—until their expedition, no man had overwintered in the middle of Greenland—and an opportunity, they hoped, to engage in "serious scientific investigation" in a region where heroic exploration and adventure reigned.[2] Despite hardships and tragedy, including the deaths of Wegener and his Greenlandic sherpa, Rasmus Villumsen, the expedition provided the first year-round data set for a central polar ice sheet location. Such was the legacy of Wegener's expedition that when French anthropologist and polar explorer Paul-Emile Victor looked to plan an expedition to Greenland immediately after World War II, he chose to pay homage to the German team by revisiting Eismitte and extending the existing scientific data. During two overwinters in 1949–1951, EPF produced emblematic scientific results pointing to ice melt and rising sea levels. For the United States, Eismitte was a laboratory of the Cold War. During that conflict, when Greenland was paramount to North American continental security, the US military took a keen scientific interest in the island. In 1955, a team of six US researchers traveled overland to Eismitte to build the scientific knowledge of ice and snow necessary for military purposes. The scientific momentum of the location meant that for the European cooperative project EGIG, Eismitte stood out as a critical stop. For the Europeans, Eismitte offered the chance to better understand atmospheric contamination and climatic change in a decade when such issues were gaining traction in the scientific world. Finally, as the introduction of intercontinental ballistic missiles in the mid 1950s made accurate targeting a priority, Eismitte played host to a US Air Force geodetic mapping station.

To stand at Eismitte today is to stand atop layers of old camps buried progressively deeper in the snow: 40 meters down are the bones of a US Air Force radar station, built to link Greenland to the outside world; just beneath lie traces of Project Jello's mobile scientific camp; another five meters down are the remnants of the French *Station Centrale* (Central Station), its prefabricated walls and laboratories buckled under the weight of accumulated snow; and nearly 60 meters below the surface is the snow cave from which Wegener set out on his last, fateful traverse, crushed almost to nothing. Through all these layers snake ice core drilling channels, now twisted and fractured but still reaching deep into the ice and into the past. Thanks to an average annual snowfall of 400 kilograms per square meter—layers of snow which never melt entirely, becoming denser as new snow falls, eventually turning into ice and moving ever downwards, feeding the great glacier—Greenland's ice sheet acts as a time capsule: to travel down in the ice is to travel back in time.[3]

At the opening of the twentieth century, the polar regions stood out as the last blank spaces on the earth's map, unknown frontiers that beckoned to be explored.[4] The Arctic and the Antarctic, and above all the poles themselves, sparked races between nations and between men. In an era when exploration was losing its heroic cachet, the polar regions fueled what Felix Driver elegantly terms "the continued imaginative appeal of exploration in the modern world."[5] This term *imagination*, so common in the polar context, provides a valuable window into scientific exploration by challenging us to see the process of discovery as a transformative one, of both man and environment.[6] The angle taken in this book—that of *scientific* imagination—encompasses the scientific questions that enticed so many men to Eismitte and the broader narratives surrounding those questions. Why did the center of Greenland's ice sheet hold such appeal? How did the practices of science shape the relationship between the men (and, indeed, all were men) who traveled to Eismitte and the ice sheet? And how did the scientific perception of Eismitte change over time as access barriers to the ice sheet fell and as climatic questions came to the fore?

"How is it possible," asks Roger Launius in his historiography of polar scientific exploration, "to understand an environment that tolerates virtually no life, one that is derived from a single substance, and one that is for the most part a single color?"[7] In the nineteenth century, he answers, the scientist's mission was to transcend the "horror and enthrallment" of the polar landscape and to tame it with order and systematicity. From a starting point of "almost total ignorance" in the early 1800s, a century later the contours of the Arctic world had been traced and were beginning to be filled in.[8] As the twentieth century brought a host of technological advances, from instruments capable of probing the upper atmosphere and drilling deep into the ice to motorized vehicles and ski-equipped aircraft that opened access to the ice, scientific work in the polar world grew in quantity, quantification, and specialization. "People rarely found riches and never discovered paradise," wrote Norwegian polar hero Fridtjof Nansen, "but we always gained more knowledge."[9] When Wegener approached Eismitte in 1930, his task was to reimagine the ice sheet as a field of work, "not a desert to be crossed as quickly as possible"—the first step in transforming the center of Greenland into a site of scientific knowledge production.[10] Over the following decades, the scientific picture that emerged from Eismitte was one of accumulated knowledge, driven by an underlying desire to build a comprehensive picture of the central ice sheet, to dissect and ultimately give life to that monochromatic environment. Guided by the diverse agendas of the scientists and their patrons, meteorology, climatology, seismology, geodesy, and above all glaciology melded together to shed light on the shape, movement, and weather patterns of the ice sheet, and to reach deep into the past through ice core drilling.

The sciences practiced at Eismitte were classic field sciences. Recent work in the history of science has embraced the field sciences, raising questions about the epistemology of field- and laboratory-based scientific knowledge, the nature and legitimacy of scientific evidence from the field, and the role and claims to authority of amateur scientists (especially in the nineteenth century).[11] These questions form part of a broader, and fascinating, debate over how *place* should be accounted for in science, or, the geography of scientific knowledge.[12] Captivating as these themes are, however, they are not my object here. Rather, like Ronald E. Doel's study of the environmental sciences in the postwar United States, I want to draw out the apparatus surrounding the science, to illuminate the technological, political, diplomatic, and institutional settings that supported scientific inquiry at Eismitte.[13] Within these settings, I emphasize the physicality of the research stations at Eismitte. As recent investigations of the spatiality of scientific research stations—that is, their architecture, anthropology, and even social and political relations—show, the relationship of research stations to their environments is intimately connected with knowledge production. The spaces in which scientists work, Raf de Bont emphasizes in his study of marine stations in Italy and France, "their physical make-up and their organization, have a tangible bearing on the knowledge that is produced there."[14] The research stations transported to and built at Eismitte over the decades—from the German snow cave to prefabricated French laboratories to giant US mobile living units—represent physical expressions of the expeditions' various raisons d'être and claims to modernity. It is the "ecologies" of stations, to use Richard Burkhardt's evocative term, comprising their physical and social organization, which mediate between the scientists and the environment under study.[15] The image of Greenland that pervades today's media—one of a fragile, delicate environment teetering on the edge of "potential climate change catastrophe"—is in part a product of these research stations. It has emerged over a period of decades, spurred by the rise of the global environmental consciousness narrative and growing concerns over climatic change.[16] By tracing the course of environmental scientific inquiry at Eismitte, I aim to shed light on the historical emergence of this image.

The world's largest island and second-largest ice body, occupying a dominant position on any northern hemisphere map, Greenland (*Kalaallit Nunaat*, in the Kalaallisut language: *land of the Kalaallit*) promised untold riches to those who ventured into its interior—not, as Nansen knew so well from his pioneering crossing of Greenland in 1888, physical riches, but the reward of knowledge. In contrast to much polar exploration (and, indeed, exploration more broadly), Greenland didn't offer territorial riches, either: there were no colonial ambitions to be fulfilled by reaching Eismitte, no dreams of raising a flag and claiming a piece of land far from home. Still, sovereignty issues were not absent from the picture. Danish sovereignty over Greenland, contested by Norway in the early twentieth century, was confirmed by the International Court of Justice in 1933.[17] Owing to Greenland's strategic importance for Allied operations during World War II and the German occupation of metropolitan Denmark, the United States invoked the Monroe Doctrine in 1941, leading to a significant buildup of US military bases and personnel in Greenland.[18] The residue of the war, combined with continued NATO-backed US military presence in Greenland during the early Cold War, forced Denmark once again to work to ensure its territorial integrity.[19] Accessing Eismitte meant gaining permission from Copenhagen, a delicate diplomatic dance that was intimately intertwined with Danish sovereignty sensibilities. The Danes themselves spearheaded no expedition to Eismitte, and Greenlanders were only peripherally (and sometimes fatally) involved with the location: even when asked by France to mount a joint expedition, Denmark declined and restricted its involvement to bureaucrats and minders. Danish participation in the cooperative European expedition of 1956–1960, too, was so reserved as to draw criticisms from international partners and even from Danish scientists and officials. The manifestations of this asymmetrical relationship open the question of how practices of geographical exploration and scientific knowledge production interact with political power and diplomacy. The *pursuit* of power through knowledge is a common thread through all the expeditions discussed here: for the Americans, scientific knowledge of the ice sheet was key to dominance in an increasingly polarized Cold War sphere, and for the French, knowledge of polar environments was prerequisite to upholding territorial claims in the Antarctic, to name but two. As for the Danes, limited commitment to Eismitte was deemed necessary to exercising power and sovereignty, in both their intellectual and territorial guises.

Through the first two-thirds of the twentieth century, scientists, technicians, and bureaucrats from seven different countries visited Eismitte on five national-based and two international collaborative expeditions. In all cases, these expeditions were supported in part or wholly by national governments and/or militaries—bodies which hoped to gain power, prestige, and knowledge from the presence of their nationals at the center of Greenland. The sheer scale of resources invested in Eismitte—transporting scientists to the middle of the ice sheet, sustaining them there, and equipping their investigations—underscores the location's perceived value. From the Weimar Republic, which contributed in excess of 150,000 marks (USD $500,000 in 2012 dollars) to Wegener's expedition in the midst of post–World War I shortages, to the French Fourth Republic, which put more than 40 million francs (USD $1.2 million in 2012 dollars) into EPF at a time when reconstruction needs dominated the country, Eismitte attracted commitments that emphasize just how closely connected the remote polar location was to domestic agendas and decisions. The complex relationship between governments, militaries, and scientists, as Mark Solovey elegantly captures in his study of US patronage for the social sciences during the Cold War, links scientific legitimacy with political (or military) authority.[20] Patronage, Solovey reminds us, is a two-way street that gives patrons a vehicle for promoting their interests while also tying individual scientists into much bigger nets. In all these respects, the actions and choices of patrons left a deep imprint at Eismitte.

It is clear by now that there is no single path, straight and blazing, leading from the individual classical heroic explorer striving out into the unknown to so-called modern scientific exploration, conducted by large teams and supported with big money and big technologies. As James R. Ryan and Simon Naylor rightly put it, "scientifically motivated exploration extended backwards in time well before 1900, while heroic exploration traditionally associated with the Victorian era extended well into the 20th century."[21] Rather than the tiresome and indeed misleading task of pigeonholing the Eismitte expeditions into perceived ages of exploration, I set out to illuminate changing practices of science in the Arctic. In doing so, two roles stand out: those of individuals and of technology. First, the science conducted at Eismitte was underpinned not only by data sheets, instruments, and government patronage, but also by the personal and professional identities of the scientists themselves. Identities, Urban Wråkberg suggests, are fundamental to "the psychology of perception with its many expectations and desires" which condition observations, reporting, and even socio-professional interactions.[22] The individuals who feature in this story—among them, a feisty but pragmatic Danish administrator trying to carve a secure place for Greenland among the competing interests of many nations, a French anthropologist and self-proclaimed modernizer of Arctic travel, and an American PhD candidate swept up in the world of Cold War military–scientific research—negotiated their identities in a transnational space that challenged their scientific, personal, and political commitments. Second, in the Arctic world, transportation technologies are frequently held up as icons of the conquered northern frontier. "The airplane," Stephen Bocking writes in his study of aviation in northern Canada, "rendered the north suddenly, and radically, accessible."[23] Indeed, transportation technologies enabled the taming of nature, the assertion of epistemic authority, and the demonstration of sovereignty across the Arctic world. For those men who ventured to Eismitte, their journeys were also defined by quieter, smaller-scale technologies that made the ice sheet livable: newfangled tinned and dehydrated foods, ultramodern insulation materials, and even long-playing gramophone records. By ushering modernity into the north, both types of technologies reached out to alter the political, scientific, and cultural landscapes of Greenland. Technology turned scientific work on the ice sheet into a routine, even commonplace, proposition—to the point where, by the 1970s, Eismitte no longer occupied a critical node in Greenland's scientific imagination.

The history of polar exploration is inseparable from questions of crafted narrative, or, as Peder Roberts puts it in his work on the European Antarctic, "self-presentation."[24] Usually tightly linked to patronage—the need to raise funds for eye wateringly expensive expeditions—the narratives crafted by explorers provide a window into their political and public personas, as well as a way of understanding how they *want* to be perceived. The "public promotion of the myth of the explorer," as Beau Riffenburgh shows, is a tantalizing mixture of propaganda, business acumen, and rivalry.[25] For the men who journeyed to Eismitte, these narratives necessarily intersect with the political contexts of their times, whether it be German renewal after World War I, French territorial interests in Antarctica, the superpower conflict of the Cold War, or European cooperation in the postwar era. In negotiating these narratives, I find great merit in Michael F. Robinson's approach to his study of Arctic exploration and American culture, in which "although the reader is occasionally brought to the Arctic regions, the real action takes place back at home."[26] In a similar vein, Roberts removes the contradiction in terms from "interesting bureaucrats" and places them front and center.[27] To understand the cultures of exploration that underlie the expeditions to Eismitte, my focus has to be dual: in the field and in the backroom sites where the expeditions were envisioned, planned, financed, and, at times, fought over. The first looks at the *what* and *how* of getting to and working at Eismitte, threads that are necessarily field- and technology-oriented, while the second looks at the driving forces behind the expeditions, the *who* and *why*, and takes place in grand Parisian offices, stateside Army testing facilities, and the somber gray edifice of Denmark's Foreign Ministry. This is hence a story not just of scientists and explorers, but also of the government officials, military planners, and diplomats who never set foot at Eismitte but who nonetheless are inseparable from the location. It is also a story not just of Eismitte, but of how the location fits into larger political, scientific, and cultural narratives.

The first chapter opens with the question that so perplexed polar explorers of the eighteenth and nineteenth centuries: behind Greenland's jagged coastal peaks and vast ice plains, was there concealed an ice-free Arctic Eden, complete with virgin woods, fertile pasturelands, and mineral riches? Following the pioneering attempts of Claus Enevold Paars, Lars Dalager, Fridtjof Nansen, and their contemporaries to penetrate the great ice sheet, I illustrate the geographic mysteries and the immense difficulties of passage that defined Greenland for so long. By the interwar years, many of these mysteries had been solved and the scope for record-setting exploration in Greenland (and, indeed, in the polar sphere more broadly) was limited. But Eismitte—at the very center of Greenland's great ice sheet—stood out as untouched and unknown. The second part of the chapter looks at the first expedition to Eismitte, led by Alfred Wegener in 1930–1931—an expedition that both triumphed in the first overwinter in the middle of the ice sheet and mourned the tragic death of Germany's foremost polar explorer. The scientific work of the Wegener expedition established Eismitte as key to unlocking Greenland's remaining secrets and set in motion decades of continued investigation.

The second chapter examines the French EPF expedition of 1948–1953, led by the dean of twentieth-century French polar exploration, Paul-Emile Victor. With its well-equipped under-ice camp at Eismitte—luxurious by the standards of the German snow cave—as well as its use of cutting-edge transportation technologies, all touted to the French public as the arrival of modernism in the Arctic, EPF has long been depicted as a technological revolutionary in polar exploration.[28] By situating the expedition in its transnational context and probing the roles of publicity and fundraising in Victor's work, I suggest that the EPF expedition to Eismitte marks not a leap from adventure to science, but rather showcases the successful crafting of a modernistic narrative. Iconic images of the expedition—from parachuting fresh melons to men camped at the center of Greenland's ice sheet to giant motorized ice tractors carefully navigating deep crevasses—were showcased in popular articles, gala evenings, children's stories, and even the Cannes Film Festival. By carefully weaving scientific and political justifications into an appealing narrative, Victor was able to position himself and his organization at the forefront of French polar work for nearly half a century.

The third chapter considers US scientific activity at Eismitte in 1955, at the height of Cold War tensions across the Arctic. Situated on the shortest great circle route between Moscow and Washington, Greenland was vital to the US military's vision of North American continental security. As part of a massive US buildup on the island, which included erecting the second-largest US Air Force base in the world more than 1,000 kilometers north of the Arctic Circle, a scientific team traveled overland to Eismitte to study snow, ice and permafrost. Led by a young doctoral student from Minnesota, Carl S. Benson, the expedition—nicknamed Project Jello—provides a lens through which to understand both the evolution of Eismitte as a place of scientific knowledge production and the culture of Cold War US activity in Greenland. This remarkable story of US military–scientific interests in Greenland is just beginning to be told, and this chapter casts a first look at Project Jello.[29]

The fourth chapter opens by looking at the 1956–1960 EGIG expedition, a cooperative venture between five European countries, including Denmark. Propelled by the changing conception of climate at the time, the EGIG expedition put environmental and climatic questions at the forefront of its scientific agenda. At Eismitte, the expedition drilled its deepest ice core, shedding new light on the circulation of contaminants through the atmosphere and even capturing a clear signature of the first Russian H-bomb tests, conducted several years earlier. But these scientific successes were marred by political infighting and the expedition ended with an aggrieved Denmark actively turning against foreign glaciological research on Greenland's ice sheet. By considering scientific knowledge production in the context of asymmetrical power relationships, I illustrate the challenges faced by a small state trying to navigate competing domestic and foreign demands. The chapter then examines Eismitte's integration into the US military's global geodetic network, designed for missile targeting at a time when intercontinental ballistic missiles were turning a new page in the Cold War. This story both ties into long-standing scientific interests at Eismitte and is representative of Eismitte's decline as a singular place of importance in the latter third of the twentieth century. With the changes that swept over Greenland's ice sheet in that era—from improved air access to permanent research stations to remote sensing technologies—the mystery and allure of Eismitte slowly waned.

Chapter 1

A Land Apart

Greenland in the Historical Imagination

On the last day of July 1878, Danish naval officer and Arctic explorer Jens Arnold Diderich Jensen cast his eyes on a jagged mountain rising improbably out of Greenland's ice sheet, its gray shape encircled by a dazzling frozen sea of white.[1] Jensen and his three companions had reached the foot of the nunatak a week earlier, utterly exhausted after a grueling 60-kilometer trek from southern Greenland's Frederikshaab Glacier. For 11 days, the party negotiated minefields of crevasses, deep azure chasms plunging down into the ice and into darkness, barely able to see through eyes wet and stinging from snow blindness. Their reward—a view into Greenland's sweeping interior—was delayed seven long days by gale force winds that whipped across the island, bringing mounds of fresh snow and confining the four men to their cramped tent. "The next morning, the weather was thankfully clear and I rose at once to the cairn, where I had an excellent view over the country," wrote Jensen: "to the east rose the immense flat extent of the ice sheet, as far as the eye could see, always higher and higher, until it merged with the sky...[To the west] a row of large dark mountain tops loomed sternly, inhibiting the progression of the ice."[2] Adrift on the ice sheet, Jensen and his men stared in wonder at the rocky nunatak which, with delicate white and yellow poppies quivering in the wind, a small bird, and some spiders, "seemed to us a paradise."[3] "The discovery of these, the ice-free peaks of subglacial mountains, faintly coated with earth, gave rise to a curious and exciting idea," wrote Laurence P. Kirwan, director of the Royal Geographical Society, nearly a century later: "Might there not be deep in the interior of Greenland not ice but stretches of cultivable land, rich agricultural land where corn and grass might grow in the midst of this desiccated field of ice?"[4]

That the interior of Greenland was such a mystery as late as the end of the nineteenth century speaks of the almost unfathomable difficulties of accessing and traveling over the ice sheet. Despite repeated attempts to penetrate the island, Hinrich Rink—the Danish geologist and colonial administrator who was considered the "final court of appeal on all questions touching Greenland"—adopted a tone of weary acceptance when he told the American Philosophical Society in March 1885 that "its interior can be considered as not yet visited by travelers."[5] Norwegian polar explorer Fridtjof Nansen concurred, noting that "the interior of that continent has remained a mystery to Europeans as well as to the Eskimos, and many are the opinions and suggestions which have been put forth as to its real nature."[6] Greenlanders feared the ice sheet, which to them was the realm of supernatural beings, and avoided journeying or hunting on it.[7] And as the nineteenth century drew to a close, European and North American explorers still struggled to understand the great island, undecided as to the extent of its ice coverage and even uncertain as to the northernmost reaches of its landmass. "Of course we all have a general idea of Greenland," explained American explorer Robert E. Peary upon his return from Greenland at turn of the century, and "yet the actual facts are so different from anything existing in lower latitudes, so entirely dissimilar from anything with which we are personally acquainted, and which we might use as a foundation from which to start our conception, that I doubt if one in ten even of the best-read has a true conception of the actuality of this great arctic island continent."[8]

The traveler who makes it to Greenland's shores first encounters a ring of craggy mountains and plunging fjords, a rocky ribbon ranging in width from eight to 130 kilometers encircling the island and forming a veritable barrier between the Arctic seas and the great ice sheet. The sounds and fjords of Greenland's coast, continued Rink in his American address, "represent the only highways; where they end, the vast glacier that covers the whole interior begins, and this has only exceptionally been ascended by travelers."[9] Rising hundreds of meters above sea level, dividing life from the lifeless ice, known from unknown, the solid ice walls of Greenland's glaciers were anthropomorphized by early travelers: "Here was a plastic, moving, semi-solid mass, obliterating life, swallowing rocks and islands, and plowing its way with irresistible march through the crust of an investing sea," wrote American explorer Elisha Kent Kane upon encountering northwestern Greenland's Humboldt Glacier in the middle of the nineteenth century.[10] Ascending onto the ice sheet, the traveler is greeted by a frozen landscape, monochromatic and silent except for the wind, and littered with dangerous crevasses. "The traveler across its frozen wastes, traveling as I have week after week," Peary continued, "sees outside of himself and his own party but three things in all the world, namely, the infinite expanse of the frozen plain, the infinite dome of the cold blue sky, and the cold white sun—nothing but these."[11]

At the very center of this enigmatic island lies Eismitte, literally *middle ice*: a location, which, although not named until 1930, pulled on the minds of explorers for centuries, distant, alien, and, for the longest time, unattainable and unknowable. No settlement, no town, not even an identifiable geographical feature marks Eismitte off from the rest of the sweeping ice sheet, a vast frozen expanse 40 times as large as metropolitan Denmark. Rather, Eismitte represents a holy grail in polar exploration: to penetrate Greenland's interior to the very middle, to stand as far from any coast as possible, to be at "the end of the world," in the words of French polar explorer Paul-Emile Victor.[12] This book is the story of Eismitte.

"Arctic Work Is Not Quite a Picnic"[13]: Penetrating the Great Island

The first European attempt to penetrate into Greenland's interior came on April 25, 1728, when Danish Major Claus Enevold Paars set out from Godthaab (now Nuuk) to cross the ice sheet with a small party of men and officers. Godthaab, on Greenland's southwestern coast, had been founded earlier that year as the new seat of the small and ill-fated Danish colony in Greenland.[14] Sent by King Frederik IV of Denmark with some two dozen soldiers, their wives and children, Paars's orders were "to cross Greenland on horseback from the west coast, and when he reached the east coast to build a fortress, found a colony, and take care of the old Norsemen who were thought still to survive."[15] After two days of trekking up from Ameralik Fjord, a long finger of water protruding into the interior just south of the colony, "we came to the ice mountain at midday on the third day," Paars wrote—but upon ascending the glacier he and his party encountered a maze of crevasses plunging deep into the ice.[16] "After advancing for a few hours in great danger of our lives all further progress became impossible," continued Paars in his vivid account of the expedition: "when we realized no further progress was possible we sat down on the ice, fired nine Danish shots from our rifles, and drank to the health of our most gracious king a glass of brandy, an honor which had never before taken place on a glacier," before turning back to Godthaab. Paars's failed mission to cross the ice sheet did, however, yield an enticing description of Greenland's interior: it is "like looking at the wild sea where no land can be seen. From the ice sheet, there is nothing to see except the sky and glittering ice...sharp-edged like white sugar candy."

The next attempt at crossing the island did not come for another quarter century. In the summer of 1751, Lars Dalager, a Danish trader based in Frederikshaab (now Paamiut), a trading post at Greenland's southern tip, was captivated upon hearing of "a new discovery made...by a Greenlander who had been so high up while out hunting that he could see distinctly, he said, the old Kablunak mountains on the eastern side."[17] Dalager immediately resolved to journey across the ice sheet to find the long-lost Østerbygd, or Eastern Settlement, established by the Norse in 985 and now thought to have disappeared during the fifteenth century.[18] At the beginning of September, Dalager—accompanied by the hunter and his daughter—ascended onto the ice sheet from a deep fjord on the southern side of Frederikshaab Glacier. Finding themselves amidst a minefield of crevasses, the party took more than seven hours to cover the first eight kilometers. At the top, the hunter pointed out ice-free peaks in the distance, the supposed eastern coast of the island. In fact, these were the same nunataks that Jensen would camp at the foot of over a century later. But Dalager and his companions never made it to the rocky mountains, forced to turn back when their footgear gave out: "although each of us was provided with two good pairs of boots for the journey, they were already worn out by the sharpness of the ice and stones [and] we were as good as barefoot," wrote Dalager after his return to Frederikshaab.[19] "I cannot here fail to mention," he added, "with what appetite I emptied a whole bottle of Portuguese wine that evening, after which I fell asleep until the following day at noon."

For a century after Dalager's journey, there was no further exploratory work on the ice sheet. In the middle of the nineteenth century, Hinrich Rink—the geologist and colonial administrator—carried out a remarkable voyage along the coast between Upernavik, midway up the western side of Greenland, and Julianehaab (now Qaqortoq), on the island's southern tip.[20] With the publication of his new geographic and geological maps of the island, Rink drew scientific attention to Greenland's great ice sheet, or, as he called it, *Indlandsisen* (the inland ice).[21] That ice sheet, Rink asserted, supplied the Arctic and North Atlantic oceans with their great icebergs—calving, by his calculations, eight to ten million cubic feet of ice every year. In Europe, Rink's description of the enormous glacial forces at work in Greenland added fuel to the ongoing debates over Ice Age theories (and, in particular, the idea that vast swaths of northern Europe had once been covered by ice). Rink's work awakened new interest in Greenland's ice sheet and set in motion a series of renewed attempts to penetrate its interior.

The three decades following the elaboration of Rink's ideas—from the 1860s until the 1880s—saw more traffic than ever before on Greenland's ice sheet.[22] In 1870, Finnish-Swedish geologist and mineralogist Adolf Erik Nordenskiöld penetrated 64 kilometers inland from Godthaab and spent seven days on the ice sheet before abandoning his attempt to cross the island at its narrow southern end. Nordenskiöld went on to gain fame by conquering the Northeast Passage, sailing from Sweden around the north coast of Eurasia to East Cape (now Cape Dezhnev), but his desire to validate his long-held belief that Greenland's interior held an ice-free Arctic Eden was never far from his mind. He returned to the ice sheet in 1883, better equipped and with a larger party of nine, and managed the furthest penetration to date, traveling 230 kilometers inland before being forced back by ferocious weather.[23] Despite seeing no signs of fertile pasturelands, his belief in an ice-free interior was unshaken. Three years later, Robert E. Peary, then a young engineer in the US Navy, claimed to have penetrated 180 kilometers inland with skis and snowshoes before turning back due to inadequate provisions.[24] With a close eye on these efforts, Rink—now retired, in failing health, and living close to his daughter in Christiania (now Oslo)—asked, "Can it be expected that Greenland once will be crossed from west to east or vice versa?"[25] He answered his own question: "I am convinced that this will be accomplished," he wrote, and he was proven true three years later when Fridtjof Nansen, then 26 years old, achieved the first crossing of the island.

Rather than starting from the Danish settlements on Greenland's western coast and heading towards the wild eastern side of the island, as all attempts to date had done, Nansen proposed to cross the island in the opposite direction. "If we started from the west coast of Greenland I was quite sure we should not be able to cross, for then we should have the flesh-pots of Egypt behind us," he recounted after the fact: by beginning from the east, all temptation to return would evaporate, the only choices being, in Nansen's words, "death or the west coast of Greenland."[26] Armed with his trusty skis (which, as a Norwegian, formed part of his national consciousness) and two months' worth of supplies packed on lightweight sledges, and joined by five companions chosen for their skiing abilities, Nansen and his party began their crossing on August 15, 1888, from Umivik, at the mouth of a long fjord due east of Godthaab. Facing problems all too familiar to those who came before him—treacherous crevasses, soft snow, and violent winds that prevented all progress for days at a time—Nansen soon altered his course, deciding to head directly for Godthaab (400 kilometers away) instead of Christianhaab (now Qasigiannguit, 550 kilometers away) as had initially been planned. After 49 days of skiing and trekking, Nansen arrived to a gunfire salute in Godthaab on October 3, 1888. Greenland had been crossed for the first time, and Nansen celebrated his achievement under a red-and-white swallow tailed Danish *Splitflag* flying high in the autumn wind.[27] Soon after his crossing, Nansen pronounced on Greenland's age-old question: far from fertile valleys and agricultural lands, "the expedition from which I have just returned has, in my opinion, fully proved...that this part of Greenland is not only ice- and snow-clad, but has a mighty shield-shaped covering of snow and ice, under which mountains as well as valleys have quite disappeared."[28] About the northern portion of the island, though, Nansen was careful to draw no conclusions.

While Nansen had conquered one of Greenland's longstanding challenges, another major challenge remained open: at the opening of the twentieth century, no man had overwintered on the ice sheet itself. In 1912, shortly before war broke out in Europe, Danish explorer and military captain Johan Peter Koch and German scientist Alfred Wegener set their sights on this goal, combining an overwinter on the ice margin near Dronning Louise Land with a traverse of the island at its broadest point the following spring. From their prefabricated hut, affectionately named *Borg* (Castle), Koch and Wegener busied themselves with glaciological, meteorological, and astronomical studies through the winter, recording some of the earliest systematic wintertime observations from the ice sheet.[29] Setting out in April 1913 with ten Icelandic ponies as draught animals, the men embarked on an ambitious 1,100-kilometer journey across the broadest section of the island. Despite a series of almost unbelievable calamities—among them, the escape of six ponies immediately upon landing in Greenland; Wegener falling and breaking a rib while reconnoitering; a calving glacier that sent a giant block of ice plummeting down onto their camp, destroying half of their supplies; Koch falling into a 15-meter deep crevasse and breaking his right leg, leaving him confined to his bed for three months during the winter; the loss of their only theodolite to the same crevasse; and large open wounds on their faces from the blistering Arctic sun reflecting off the snow—they reached the west coast by midsummer. All of the ponies succumbed to cold, exhaustion, and snow blindness, and had to be slaughtered over the course of the traverse—and when provisions ran out in the final days, the expedition dog was shot and cooked.[30] Their path, well north of Nansen's pioneering traverse, finally put to rest the myth of a green, ice-free area in central Greenland. Emboldened by the trip, the German explorer Wegener was eager to return to the ice sheet to conduct more scientific work and, in particular, to overwinter at the center of the island—but the interruptions of World War I and the subsequent upheaval in his homeland meant that his dream of reaching Eismitte would have to wait almost two decades.

Greenland as a Strategic Geopolitical Space

From the time of these early expeditions into Greenland's interior, and even before, Greenland beckoned not only to explorers and polar enthusiasts, but also to kings, governments, militaries, colonists, and merchants. Positioned between the upper reaches of the Eurasian and North American continents, the island has long been a strategic space for trade, for nation building, and for war—even though it was not in fact known to *be* an island until the turn of the century.[31]

Greenland's earliest links with the outside world date from the prehistoric times, when, around 2,500 BC, Paleo-Eskimos—including the Saqqaq and Independence I cultures—migrated from what is now Canada's Arctic archipelago to Greenland's northern tip.[32] Over the following centuries, cultures arrived and disappeared, leaving settlement traces including stone and whalebone tools along Greenland's extensive coastlines.[33] Still from the west, the ancestors of today's *Kalaallit*, the Thule culture, crossed over to Greenland around 1,300 AD. And from the east, Europeans became aware of Greenland even before the Thule arrived, when Gunnbjørn Ulfsson was blown off course while sailing from Norway to Iceland in the early tenth century. When Norse seafarers reached Greenland's southern coast at the end of that century, they found deep fjords lined with grassy fields and stands of alder and birch. There they established two settlements, raising goats, sheep, and cattle, and hunting seals.[34] According to the sagas of the Icelanders, Erik the Red named the land *Grænland*, because, he said, "men will desire much the more to go there if the land has a good name."[35] The Norse settlements linked Greenland to Europe through regular trade: salt, timber, and iron tools came in on trade vessels from northern Europe, while the Greenland settlers exported valuable polar bear furs, seal skin rope, and walrus and narwhale tusks. When the settlements joined the Kingdom of Norway in 1261, Bergen imposed strict controls over this trade, marking the beginning of European claims to authority over the island. Greenland's Norse settlements, however, started to decline in the fourteenth century and disappeared by the middle of the fifteenth century, for reasons still not clearly understood.[36]

The race to discover a Northwest Passage to the Orient put Greenland on the route of grand exploratory expeditions in the late fifteenth and sixteenth centuries. As Danish, English, and Portuguese explorers sailed across the high North Atlantic, Greenland's southern coastline gradually solidified on maps.[37] Through this era, Greenland was thought to form part of a sweeping polar continent stretching from Eurasia to the North American Arctic. Maps from the sixteenth and seventeenth centuries, such as that produced by Icelander Sigurdur Stefánsson in c. 1590, show Greenland as part of the polar mainland, a long oak leaf shaped peninsula jutting southwards midway between Eurasia and North America—and making *Grönlandia*, as Stefánsson labelled the land, central to what historian Kirsten A. Seaver calls "the greed of America-fever."[38]

The land between the continents was not only of interest to those making their way to the New World (and, many hoped, onwards to the Orient): it was also a vast and sparsely populated space in which the Nordic realms—namely, Denmark and Norway (and their various incarnations in the modern era)—played out their territorial ambitions. The efforts of Claus Enevold Paars and Lars Dalager to penetrate into Greenland's interior in the first half of the eighteenth century form part of Greenland's many-layered colonization history—a history which centers on competing national interests in the island and underscores centuries of both cooperation and conflict within *Norden*. From the integration of Greenland, then a Norwegian dependency, into the personal union of Denmark-Norway in 1536 to the final International Court of Justice ruling in 1933 that granted Denmark sovereignty over all of Greenland, the island sat at an uneasy crossroads spanning the Arctic and high North Atlantic oceans, of strategic importance to Danish colonists, Dutch whalers, and Norwegian merchants, among others.

In the nineteenth century, Greenland began to attract interest from across the Atlantic as explorers, businessmen, and politicians from the United States cast their sights on the island.[39] Robert E. Peary's territorial claims on behalf of the United States, made during the nearly four years he spent in Greenland between 1886 and 1897, were only resolved in 1916 with a treaty that saw the United States relinquish

Peary's claims in return for Denmark agreeing to sell the Danish West Indies (they are now known as the United States Virgin Islands).[40] And in the late nineteenth and early twentieth centuries, Greenland attracted US political attention as a site for naval and aviation installations and US commercial interest in the island's coal, cryolite, and hydroelectric potential.

Greenland took on global strategic significance during World War II. With the opening of the war, the island's position as a stepping-stone between the continents drew the attention of all parties: to US military strategists, Greenland was an empty space in a critical region requiring US presence to prevent the Axis powers from securing the island; to the Germans, it was a source of much-desired meteorological data for U-boat campaigns and for planning air attacks over Europe; and to the Danes, the great island was too much to defend. Following the German occupation of metropolitan Denmark in April 1940 and the ensuing severing of connections between Denmark and the colonial authorities in Greenland, President Franklin D. Roosevelt—with the agreement of the Danish representative in Washington, Henrik Kauffman—allocated five million dollars to construct US military bases in Greenland.[41] The opening of US military presence on the island came in 1941, when Norwegian-born Colonel Bernt Balchen's battalion was tasked to build an airfield at Søndre Strømfjord (codenamed Bluie West 8) to act as a ferrying station for bombers en route between North America and Europe.[42] Designed to protect transatlantic communication lines, to provide a stepping-stone for aircraft traveling between the continents, and to prevent German forces from setting up shop in Greenland, the Søndre Strømfjord airfield and other US bases were augmented by meteorological stations which provided important weather forecasts for Allied operations.

During World War II, US presence in Greenland was governed by the 1941 bilateral Agreement Relating to the Defense of Greenland, which granted the United States open access to Greenland "until it is agreed that the present dangers to the peace and security of the American continent have passed."[43] At the end of the war, the Danish government reasserted some control by taking over US meteorological stations on the island, but the United States maintained the airfield at Søndre Strømfjord. Viewing Greenland as a strategically vital region in the emerging Cold War, the United States went so far as to make a formal offer to purchase the island.[44] Danish Foreign Minister Gustav Rasmussen immediately rebuffed this offer, but this did not mark the end of US presence in Greenland. In a move which broke with Denmark's past traditions of neutrality and nonalignment, in

April 1949 the small country joined the North Atlantic Treaty.[45] The question of Greenland's defense was drawn into the North Atlantic Treaty Organization (NATO) sphere and, in 1951, a new NATO-backed Defense of Greenland Agreement gave the United States the right to exercise exclusive jurisdiction over three defense areas on the island.[46] Located at Thule, in northeast Greenland, Narssarsuaq, on the island's southern tip, and Søndre Strømfjord, two thirds of the way down the western coast, the defense areas became the mainstays of US activity in Greenland through the Cold War.

Within a year of the 1951 agreement, Greenland was drawn concretely into the superpower conflict when the United States erected Thule Air Base on the island's northwest coast and began a "massive but restricted and controlled military colonization."[47] Located midway beneath the trajectory most likely to be followed by any missiles launched by the superpowers, Thule was the primary US operations base in Greenland through the Cold War. At the time of its construction, Thule was the northernmost US Air Force base, and at 90,000 acres it was second in size only to the Strategic Air Command Headquarters in Nebraska. Thule was initially designed to enable US B-47 bombers to reach Soviet industrial targets with one airborne refueling—a strategic necessity in the years before the development of intercontinental bombers.[48] By the 1960s, Greenland's position in the Cold War had changed as new technologies and a reevaluation of US strategy transformed Thule into an early warning site: the Ballistic Missile Early Warning System installed in 1958–1959 tied Greenland to the North American Air Defense Command (NORAD) Center inside Colorado's Cheyenne Mountain and NATO's Supreme Headquarters Allied Powers Europe (SHAPE) in Belgium, which together aimed to give US planners 15 minutes notice of a Soviet attack. In all of these aspects, Greenland represented a critical Arctic node of North American continental defense—an aspect of the island's strategic geopolitical position taken up further in chapter 3.[49] For now, however, we return to the interwar years and Alfred Wegener's journey to the center of Greenland.

To Eismitte: Alfred Wegener's Final Expedition[50]

Germany in the Polar World

Alfred Wegener's final expedition—an expedition that would see both the first overwinter in the very middle of Greenland and Wegener's death on the ice sheet—had its roots in German polar exploration of the late nineteenth and early twentieth centuries. While the German explorers of that era—Erich Dagobert von Drygalski, Wilhelm Filchner, Karl Koldewey and their contemporaries—do not conjure up the same heroic, triumphant image as those Britons, Americans and Scandinavians who accomplished the "polar firsts," still they were part of a national polar agenda driven by geopolitical design, scientific ambition, and geographical curiosity. Von Drygalski's Antarctic expedition of 1902–1903, for example, was supported at home by a political culture that saw a south polar expedition as an important signifier of German unification, national prestige, and scientific responsibility. In the aftermath of World War I, however, with the collapse of Imperial Germany and signing of the Treaty of Versailles, the weak and tumultuous Weimar Republic no longer had the confidence or means to reascend into the polar regions. "Forced by the peace settlement to disband German armies, surrender German territory, and deliver a significant portion of the German people into foreign rule," writes David Thomas Murphy in his survey of German polar activity, "the age of German aspiration to world-shaking deeds, among them polar exploration, seemed to have passed for good."[51]

Still, as years of hyperinflation gave way to the *Goldene Zwanziger* (Golden Twenties), which saw economic growth and declining civil disorder, polar-minded scientists and politicians began to speak anew of a German return to the polar regions. To Friedrich Schmidt-Ott, acting vice president of the *Kaiser-Wilhelm-Gesellschaft zur Förderung der Wissenschaften* (Kaiser Wilhelm Society for the Advancement of Science) and cofounder of the *Notgemeinschaft der Deutschen Wissenschaft* (Emergency Association of German Science) and likeminded colleagues, the polar regions were "more vital than ever, providing a site where German technology, science, and discipline show the world that, even in defeat, Germany remained a power of global importance."[52] This renewal of interest was driven by a desire to maintain scientific, technological, and industrial prestige in face of growing international competition. In this respect, two elements stand out: German-developed seismic instruments for ice thickness measurements and propeller sleds for polar transport. "Delay creates the danger that foreigners will precede us in the use of ice depth measurements on the inland ice, and that the new insights that are to be anticipated precisely in the first application of this new research method will be lost to German research," wrote German meteorologist and polar enthusiast Wilhelm Meinardus in 1929: "Likewise, it would naturally be regrettable if by postponement the first attempt to use propeller sleds on the inland ice came from another venue, and

German technology would suffer a loss of prestige if we are not first to have our cargo driven over the inland ice by Siemens motors."[53] For Schmidt-Ott, too, these new polar technologies were critical to ensuring that "Germany should not be shut out of the international competition of the nations in polar exploration."[54]

In this climate, an ambitious expedition to Greenland began to take shape. Spearheaded by German meteorologist and old Greenland hand Alfred Wegener, the expedition (known in German as *Die Deutsche Grönland-Expedition Alfred Wegener*, and referred to here as "the Wegener expedition") was designed to be the first serious scientific expedition to the ice sheet, anchored by an overwinter station at the very center of the island—a feat never before attempted. With a corps of nearly two dozen scientists—far more than most polar expeditions of the era—Wegener aimed to place Germany at the head of a "new style of exploration which involves the intensive cult of physical science at fixed stations as opposed to the old method of casual journeys through unknown country."[55] The expedition was a collaborative effort between the old guard of German polar research and sympathetic scientists and government officials who by the late 1920s were united in the "search for polar redemption."[56] Schmidt-Ott secured an initial commitment of 150,000 marks (USD $500,000 in 2012 dollars) from the Emergency Association of German Science,[57] and in the following years the association continued to cover the expedition's ever rising costs—an extraordinary commitment given the domestic situation: "The winter of 1929–30, in the aftermath of the global economic crisis initiated in October on the New York Stock Exchange, was a moment of unparalleled financial need for the Weimar government. Despite their desperate straits, however, scientific ideals, nationalism, and cultural values convinced Weimar officials to back the Greenland expedition."[58] With this generous commitment, as Swedish glaciologist Hans W. Ahlmann wrote admiringly in 1941, Wegener had "as good and complete a technical and scientific equipment as possible," allowing for an "exceedingly comprehensive" program of work.[59]

Alfred Wegener

Remembered as "the dominant historical figure presaging the plate tectonics revolution" and "undoubtedly the most heroic figure in the annals of German exploration," Alfred Wegener's name conjures up images of heroism, tragedy, and a man ahead of his time.[60] Born in Berlin in 1880, Wegener was an avid outdoorsman from a young age, enjoying hiking, sailing, and mountaineering with his four older siblings. His education in Berlin and Heidelberg covered physics, mathematics, and astronomy, and included a yearlong assistantship in 1902–1903 at Berlin's Urania observatory, one of the first major European centers to promote science to a lay audience. Two years after finishing at Urania, Wegener completed his doctorate in astronomy at Friedrich-Wilhelms-Universität (today the Humboldt University of Berlin). Together with his elder brother Kurt, he then joined the new *Königlich-Preußische Aeronautische Observatorium* (Royal Prussian Aeronautical Observatory). The brothers conducted meteorology and ballooning research as part of the new science of aerology (that is, the meteorological study of the total vertical extent of the atmosphere), pioneering the use of weather balloons to track the movement of air masses. In April 1906, as part of a meteorological and celestial navigation study, Alfred and Kurt Wegener remained aloft in their balloon for more than 52 hours, setting a new world record for the longest balloon flight. It was this meteorological experience that brought Wegener to Greenland for the first time, as part of the Danish *Danmark* expedition of 1906–1908.

Led by Danish explorer and ethnographer Ludvig Mylius-Erichsen, the *Danmark* expedition set out to map the last unknown section of Greenland's coastline, in the very northeast of the island, and in doing so to demonstrate Danish sovereignty over northernmost Greenland. With a team of 27 men, including two artists who struggled to keep their oil paints from freezing in temperatures which plunged below -40°C, the expedition covered more than 6,000 kilometers by dog sled and proved existing theories about Greenland's northern landmass incorrect—but at great cost: Mylius-Erichsen and two of his companions died of starvation and exhaustion on a sled journey gone terribly wrong.[61] As a junior member of the expedition, Wegener was far removed from this tragedy, spending most of the two years *in situ* at the expedition's base—the ship *Danmark*—conducting meteorological and astronomical observations. Wegener occupied the long winter days by improving his command of Danish and meticulously laying out future polar work in his diary, planned together with the expedition's cartographer, Johan Peter Koch. Between reading Danish writer J. P. Jacobsen and discussing the technical nuances of ice photography with Koch, Wegener soon overcame the language barrier and, by the end of the expedition, was comfortable in Danish. But he was not satisfied with his junior position and spent the two years "living here absorbed in thoughts of a future German expedition" in which he would play a leading role.[62] This dream came partly to fruition four years later when, in 1912–1913, Wegener returned to Greenland with Koch and overwintered on the edge of the ice sheet before crossing the island from east to west.

The years leading up to World War I were a busy time for Wegener: as well as his audacious expedition with Koch, he composed his "daring hypothesis" about continental drift in 1911.[63] In opposition to the reigning view of the era—that the oceans were permanent and unchanging features of the earth—Wegener proposed that the continents had begun as a single landmass, slowly breaking apart and drifting into their present-day formation. Early in 1912, he presented his ideas to the annual meeting of the German geological association, *Geologische Vereinigung*, in Frankfurt, and then published them in the acclaimed journal *Petermanns Geographische Mitteilungen*, and later as a book titled *Die Entstehung der Kontinente und Ozeane* (The Origin of the Continents and Oceans).[64] His hypothesis was greeted with derision, and continental drift was not widely accepted until well after Wegener's death, when ocean floor explorations in the 1950s and 1960s cast new light on the earth's crust.[65] During the war itself, Wegener was shot twice at the Battle of Belgium before being discharged to the field meteorological service. The war years also saw Wegener's marriage to Else Köppen, the daughter of his mentor in meteorology, Wladimir Köppen.

The war over, Wegener first succeeded his father-in-law as chief of the meteorological branch of the German naval observatory, *Deutsche Seewarte*, in Hamburg, and then, in 1924, finally obtained his desired professorship in geophysics and meteorology at the University of Graz. Throughout these years, a return to Greenland tugged quietly at the back of Wegener's mind, but between the political and economic realities of the post–World War I era and his Danish colleague Koch's illness and death, a return to the polar regions seemed a near impossibility. By the late 1920s, however, reemerging German interest in the polar world breathed life into Wegener's hopes. Between Wegener's previous experience in Greenland and the specific interests of leading German scientific and political figures, a "boldly and rationally designed" expedition came into being.[66]

When Wegener began to plan his expedition, central Greenland stood out as one of the last great unexplored regions of the world. "There was...little scope left in the Arctic in the second quarter of the 20th century for record-breaking journeys of [great] magnitude," wrote the Royal Geographical Society's Laurence P. Kirwan: "Indeed only within the great ice sheet covering Greenland was there scope for exploration on anything like a grand scale."[67] By the late 1920s,

Greenland's contours had been mapped, its fjords, inlets and major coastal glaciers prodded and examined, and the island had been criss-crossed at several latitudes. That Greenland did not, in fact, harbor green oases was no longer in doubt. The focus of exploration shifted from crossing over the ice sheet to living on the ice sheet, from summer treks to winter sojourns. The new Holy Grail was to overwinter at the center of the island, as far from the coasts as possible—and as far from rescue as could be. To Wegener, overwintering in the middle of Greenland also held tremendous scientific promise: the ideal expedition for a man who, nearly a quarter of a century previously, had so fastidiously laid out plans for a major German polar expedition in his diary.

Wegener's aim was ambitious: he wanted to establish three scientific stations in Greenland, one each on the eastern and western coasts and one at the center of the ice sheet—a location so hostile and remote that it had seen no previous overwinters. By manning the stations through the entire year of 1930–1931, Wegener hoped to collect the first comprehensive set of glaciological and meteorological data from a full cycle of the seasons on the ice sheet. Of chief interest to this chapter, and indeed "the main object of the expedition" (in Wegener's words), is the central ice sheet station, situated at 71° 8' North, 40° 3' West—a location that Wegener named *Eismitte*, or *middle ice*.[68] Set 400 kilometers from Greenland's eastern and western coasts and at an elevation of 3,000 meters above sea level, Eismitte represented the remotest location in all of Greenland.[69]

The selection of Greenland as the focus of Germany's return to the polar world was facilitated by Wegener's strong ties to Denmark—ties that had been built up during his two pre–World War I expeditions to the Danish colony. Between his close friendship with former expedition-mate Johan Peter Koch and his fluent Danish, perfected over three years of work in Greenland, Wegener was considered an honorary Dane.[70] The Danish authorities quickly approved his expedition and granted permission for the Germans to import fuel and other prohibited items into Greenland.[71] Denmark also provided logistical assistance in the form of two ships, the Danish government's *Gustav Holm* and the Greenland Office's *Disko*, which transported the German team to Greenland's western coast in the spring of 1930.[72] No Danish scientist participated in the expedition—Wegener's old companion Koch had died two years before—but Peter Freuchen, an archaeologist and old Greenland hand, and Helge Larsen, a young assistant at the Danish National Museum's ethnographic collection, accompanied the Germans on early trial propeller sledge runs.[73]

Establishing Station Eismitte

Writing in 1941, Swedish glaciologist Hans W. Ahlmann described the keystone of Wegener's expedition, and ultimately the cause of Wegener's death, bluntly and succinctly: "The whole enterprise depended for its success on the establishment of the Eismitte station—3,030 meters above the sea and 400 kilometers from the west coast—sufficiently early in the autumn of 1930. This was a very difficult task, which was not successfully accomplished, the difficulties proving too great.... It was the belated establishment of the station, in conjunction with other more personal factors, that cost Alfred Wegener his life."[74] The establishment of Station Eismitte in the summer and autumn of 1930 saw careful planning unravel as the weather refused to cooperate and much-touted new technologies performed poorly on the ice sheet—a drama that banded together Wegener and his three fellow countrymen who overwintered at Eismitte: Johannes Georgi, Fritz Loewe, and Ernst Sorge.

Georgi's and Wegener's paths had crossed many times before, when Georgi studied meteorology in Marburg immediately after World War I and, later, as colleagues in the meteorological division of Hamburg's Naval Observatory. In the mid-1920s, Georgi traveled to the northern tip of Iceland to probe the higher layers of the atmosphere with weather balloons, where he unexpectedly encountered jet streams.[75] This perplexing discovery, together with his first taste of Greenland, whose shores he reached in 1928 aboard the German research vessel *Meteor*, whetted his appetite for a major meteorological expedition to Greenland. How, Georgi wanted to know, did Greenland's great ice sheet influence weather patterns in the North Atlantic?—a question that fit neatly into Wegener's plans. Loewe's path to Eismitte was more roundabout: his doctorate, awarded in Berlin in 1924, examined the geographic distribution of rainfall in Africa. His dreams of being a pilot dashed by his poor eyesight, Loewe joined the Royal Prussian Aeronautical Observatory, following in the footsteps of both Wegener brothers, before being invited on the Greenland expedition. Sorge's academic roots, too, lay in warmer climes, with a dissertation on South America's dry border regions. It was his experience with glaciers—including research in Iceland and the Alpine mountains in the 1920s—that lifted Sorge out of his position as a geography teacher in Berlin's school system and into the fold of the Wegener expedition.

To set up and supply Station Eismitte, Wegener's team needed to transport four tons of supplies over 400 kilometers of barren ice from their base on Greenland's western coast to the center of the island. While Wegener had planned to make several propeller sledge and dogsled trips to Eismitte, traveling across the ice sheet proved more arduous and more challenging than anticipated. "The program of our expedition is gradually becoming seriously endangered by the stubbornness of the ice," confided Wegener in his diary in June 1930 while waiting impatiently for the winter sea ice to break up: "What we could do here is too pitifully little, and we're too few as well. And time is running out—it's already the middle of June! The affair is developing catastrophically."[76] Between scaling the steep coastal cliffs, navigating around dangerous crevasses in the marginal zone, and coaxing the expedition's Icelandic ponies into action, it took weeks to haul the propeller sledges up onto the ice sheet—by which time it was September, and the unwelcome arrival of winter weather made it all but impossible to supply Station Eismitte properly. "Only God knows why we have the worst luck with weather," continued Wegener: "The chance that our Station Eismitte will fail is embarrassingly large."[77]

Georgi and Sorge were already at Eismitte, having traveled to the center of the ice sheet by dogsled earlier in the summer. They had been relying on resupply before winter set in, and by autumn found themselves with little in the way of food and fuel. With no radio, no building materials, and few of the explosives necessary for the seismic measurements that had been so integral to the expedition's fruition, the chances of a successful overwinter were slim. "We waited and waited for the fourth sledge party which was to bring us our winter hut, paraffin and instruments," wrote Sorge, "but no one came."[78] Still, driven by the knowledge that a station at the center of the ice sheet was integral to Wegener's vision, Sorge and Georgi decided to push ahead and began excavating a snow cave to serve as living quarters and a basic scientific workshop—a cave that became known as Station Eismitte.

In a much analyzed decision, Wegener—together with Fritz Loewe and 12 Greenlanders—set out on September 21 for Eismitte in a last-ditch effort to save the central station and make the expedition a success.[79] "I am seriously concerned about the central firn station," wrote Wegener in his diary in early August, when he was still directing work at the western coastal station: "Georgi's sled trip only brought in 700 kilograms, some of which Georgi, who is staying there, has unfortunately already used."[80] From the outset, the autumn journey was fraught with problems and bad weather: as temperatures plunged below -40°C and the sleds sunk into deep, soft snow, the Greenlanders began to turn back. "The whole business is a big catastrophe and there is no use in concealing the fact," penned Wegener in a letter to Karl Weiken, the German surveyor he had left in charge of the western station, on September 28, after only a week of traveling: "it is now a matter of life and death."[81] "If only Georgi and Sorge had brought fewer instruments, more food, and their house!" he continued in a desperate tone.[82] By October 7, only three members of the party remained: Wegener, Loewe, and a young Greenlander, Rasmus Villumsen. Delayed by fierce winds and snowstorms which forced them to ride out entire days inside their tents, Wegener ordered the jettisoning of all supplies and the mission turned from one of rescue to one of survival.

The trio arrived at Eismitte on October 30, after more than five grueling weeks of travel. Greeted with firm handshakes from a much relieved Georgi and Sorge, the three voyagers stumbled into the under-snow Station Eismitte, where they drew back their tightly cinched hoods and revived themselves with chocolate proffered by their colleagues. But their respite wasn't to be for long: with no new supplies, Station Eismitte could support a maximum of three men for the winter's duration. His feet ravaged by frostbite, Loewe was in no condition to return to the coastal camp, and so settled in for the winter with Georgi and Sorge while Wegener and Villumsen embarked on their return journey. "I felt rather done-up and a few hours later crawled into my sleeping-bag, in which I was to spend over six months," wrote Loewe soon after his arrival at Eismitte.[83] Loewe's sleeping bag confinement stemmed from the amputation of his toes, which were lost one by one to incipient blood poisoning from frostbite: "Georgi cut away the flesh round the roots of the toes with his sharp knife, nipped off the bones...with a metal-cutting shears, and cut through the very sensitive big toe at the softest part," recalled Sorge: "My job was to hold the electric torch and use all my strength to hold Loewe's leg still."[84]

After posing for a photograph—the two men side by side, Villumsen a full head shorter than Wegener, both bundled into their furs and staring solemnly at the camera, with dogs curled up by their feet—Wegener and Villumsen set out on their return journey on November 1, a day which also marked Wegener's 50th birthday, and only two days after reaching Eismitte. Their attempt to reach the western coastal station at the opening of the Arctic winter, when temperatures plummeted below -50°C and days provided only a few hours of light, has an epic status among polar journeys. Wegener collapsed first, just over halfway to the coast, "probably not by freezing to death but by weakness of his heart," and Villumsen carefully sewed his body into sleeping bags and marked his grave with Wegener's own skis.[85] Villumsen continued on alone, but perished before reaching the western station, his body lost to the snow. Back at Station Eismitte, the three overwinters were isolated without a radio and did not find out that their leader and his sherpa had failed to reach their destination until the following spring. "As we watched the two men disappearing," wrote Sorge, "we did not imagine that it was the last time we should see them alive."[86] "We were cut off from the world for half a year," he continued, "and should have to rely on ourselves and the equipment which lay within 20 yards around us." At the western station, Weiken and his men assumed that Wegener and Villumsen had decided to stay the winter at Eismitte, and their fates were not discovered until the spring. Following Alfred Wegener's death, his brother Kurt came to Greenland to assume leadership of the expedition.

Transportation and Technological Failure

The failure to properly equip Station Eismitte—a failure that culminated in Wegener's and Villumsen's deaths in the early winter of 1930—is tightly linked to transportation technologies.[87] Immersed in a narrative of a new approach to scientific work in the polar world, it was clear from the outset that Wegener's expedition would not rely solely on dogs and ponies to move across the ice sheet. In the decade leading up to the Wegener expedition, aircraft had made important inroads in the polar regions: in the 1920s, US aviator and explorer Richard Byrd undertook his controversial flight near the North Pole and the Canadian government used aerial surveying and mapping to assert sovereignty over the country's north. Soon thereafter, Danish geologist and Arctic explorer Lauge Koch flew air reconnaissance missions over northeast Greenland and US aviator Charles Lindbergh flew survey flights over the island's western side on behalf of Pan American Airways. But Wegener had little time for airplanes: financially, they were prohibitively expensive; technically, flying over, landing on and taking off from the ice sheet were dangerous propositions; and morally and psychologically, Wegener "feared that his scientists would lose their motivation when they compared their strong efforts on the ground with the endless possibilities of the airplanes above."[88] Instead, Wegener chose propeller sledges, or *Propellerschlitten*. Dark against the white snow, bubble-shaped in front and tapering to a narrow tail, and with a large propeller mounted on the back, the sledges had a futuristic air to them. "The advantage of the sledges is that there is no danger in starting and stopping, as there is for an aeroplane," wrote Sorge: "Besides, they can stand for a long time in the same place on the inland ice without needing any fuel; whereas dog-sledges continuously consume food for dogs during a stay on the inland ice."[89]

Wegener's initial enthusiasm for the propeller sledges jumps off his diary pages. "We begin a new epoch in polar exploration," he declared: "What we are doing here points the way at once for future exploration. How wonderful that it should fall to us to make this pioneering step; nay, in view of the many air disasters which have occurred in polar regions, I may say this redeeming step."[90] After listening to the rumbling of the motors on the ice sheet, he continued: "I had the feeling that a dream was coming true! I am proud of the propeller sleds, because they represent a major advance in polar research...It is amazing to see that we have brought the right means of transportation!"[91] In a ceremony featuring speeches and snowballs in place of champagne, the German team baptized the sleds *Schneespatz* and *Eisbär* (Snow Sparrow and Polar Bear). Wegener's enthusiasm grew when he discovered the luxury of riding in comfort while smoking the pipe that was rarely far from his lips: "Yes, now the dream has become reality. I am comfortably seated in an enclosed cabin smoking my pipe—we have finally arrived on the ice sheet. It's an outrageous luxury. It still seems totally unreal to me."

But this early enthusiasm was dashed when it became clear that the propeller sleds could not cope with the ice sheet conditions. As problems began to rear their heads, Wegener initially remained relatively upbeat: "We can only travel efficiently with the propeller sleds when the path and wind conditions are favorable," he confided to his diary.[92] When it became apparent that the sleds were helpless in crevasse zones and soft snow, though, Wegener's tone turned to frustration, and then to desperation: "Progress is always very difficult, and it indicates that the motors are just too weak for the large and heavy sleds," he wrote in early September—and then, just a day later, "What now? Disaster has struck. It is impossible to equip the central firn station in the way we had planned." The propeller sleds never made it to Eismitte that year—a failure that ultimately cost Wegener his life.

"Twelve Months of Inland Ice Do Wear a Man Out"[93]: Overwintering at Station Eismitte

After Wegener and Villumsen's departure from Eismitte on November 1, 1930, the three remaining men—Georgi, Sorge and Loewe—were isolated at the center of Greenland's ice sheet in the German station, little more than a dugout in the snow, for 190 days, a quarter of them under polar darkness, until a relief party arrived in early May of the following year. Excavating the station—digging down by hand and with shovels, cutting unwieldy snow blocks with a small saw, and removing four tons of snow to the surface—was grueling work. The men approached their task methodically and with good-natured rivalry: "My digging experience had shown that only 6 to 9 feet down the ice was very firm and strong," wrote Georgi: "We accordingly proposed to excavate a fairly large space and erect our summer tent in it, far below the surface.... Now began a spell of rivalry in work; while I constructed my balloon-room in an underground store-room, Sorge excavated a large cave in which to set up his seismographs for the measurements of the thickness of the ice, and made observations of the temperature and density of the ice."[94] The first room to be excavated was designated for the irreplaceable barometer, which needed to be insulated from temperature fluctuations and sheltered from possible damage. Eager to begin the scientific work which would define their winter, the two men regularly paused from digging to take temperature measurements and to set up scientific instruments. While the station itself was buried inside the ice sheet, the men also built a three-meter tall snow tower for launching and tracking weather balloons. "The whole thing now looked very like a baronial castle," wrote Georgi: "my companions even likened it to the Castle of the Holy Grail in Parsifal, and jokingly addressed a letter to me at 'Castle of the Holy Grail, dungeon on the left', in reference to the underground apartments."[95]

From the surface of the ice sheet, all that showed of Station Eismitte was a low snow-brick wall fronted by the meteorological tower, cylindrical and out of place against the flat white expanse. Nearby stood a wooden weather station, held in place against the strong ice sheet winds by guide wires. Eighteen snow steps led down to a small living room with just enough space for a table and three bunks (in fact, elevated ice platforms covered with reindeer skins and waterproof sheeting), as well as separate rooms for storage, the barometer, and the weather balloons. Between the three men, their supplies, and a potpourri of scientific instruments, the ten square meters of living space were cluttered and cramped. Large fur mittens hung to dry from hooks in the walls, a coffee pot dangled from the ceiling, and all available surfaces were littered with drinking cups, oil lamps, and observation notebooks. Ensconced inside for the winter,

> our first and strongest impression was that we were lying in state in a crypt. Everything was white like marble, and our sleeping-places were clean-cut and rectangular like the marble base of a sarcophagus. The

last traces of daylight filtered down through the ceiling of ice as a mysterious blue light. In addition we had the dim light of a small lamp which lit up the vault in a ghostly, unreal way, so that it was only gradually that one realized the extent of the space. All this gave a mysterious and rather uncanny impression. Soon, however, we felt at home in the comfortable room, where no wind penetrated. The small oil-lamp worked very well; Georgi made it out of an old tin and glass photographic plates. No one was ever so well acquainted with the firn as we were. If we needed ice to make hot water, we cut a piece from the wall. As this process was repeated we at the same time acquired convenient cupboards.[96]

As new snow accumulated outside over the course of the winter and the weight over the station increased, the men installed a pillar of saw-cut snow blocks to support the station's ceiling. "Due to the heavy snow load of about 1,600 kilograms, the ceiling of our living space came down significantly," wrote Sorge, but, carefully monitored by an ingenious detection system and supported by the new pillar, it did not collapse.[97] Still, the station's exit had to be continually elongated to keep pace with the snowfall, which deposited more than three meters of fresh snow over the winter. By the time the relief party arrived in the spring, all that was visible of Station Eismitte was the top of the meteorological tower, now barely rising above the surface, and a line of wind-whipped flags, dark against the white snow.

Inside the station, the men were, in Sorge's words, "very much in the same position as men in a dugout at the front during the War. We lived in much the same way, we were just as dirty and greasy, we had lice, and we were uncertain whether we would get out of it alive."[98] The lack of supplies meant that the men had enough paraffin for cooking and lighting, but not for heating, sending them crawling into their reindeer sleeping bags during the day to keep warm. "One day is very much like another," wrote Georgi, a stoical sentiment echoed in Sorge's remark that "during the hibernation there is plenty of time to ponder the processes of nature all around you."[99] Still, the physical and mental discomforts, the monotony of the days, and the constant chilliness of their living quarters were mitigated by a rigorous scientific schedule as well as diversions including a small selection of books, holiday festivities, and the compilation of a handwritten newspaper named *Indlandsisen: En Avis for Rejsende til Grønland* (The Inland Ice: A Newspaper for Travelers to Greenland). "Yesterday another very lovely Sunday," wrote Georgi in his diary: "Sorge made coffee and observations in the morning; we conversed for hours over philosophical questions on Schopenhauer. Afternoon evaluated readings, a very fine dinner (a sort of rissole), and evening chocolate."[100] As well as daily meteorological observations and glaciological work, Sorge added, "We also discussed Schopenhauer, optimism and pessimism, religion, the growth of human consciousness, war and peace among the nations, and the probable changes in the world outside while we were wintering here."[101] This camaraderie was punctured, though, by the absence of a radio, "Our chief regret...without which we had no means of informing our companions the position we were in."

Breakfast at Station Eismitte was prepared by Georgi, the earliest riser, whose variations on porridge were fodder for one of Sorge's many vivid accounts of life at the center of Greenland: after melting ice on the stove,

> Georgi made one of his famous brands of porridge, which were never repeated—and perhaps never will be. In the course of time we had porridge with apricots, porridge with prunes, porridge with lemon drops, porridge with crystallized citric acid, porridge with chocolate, porridge with coffee, porridge with soup cubes, porridge with leftover bread, porridge with brawn, porridge with melted butter and sugar, porridge with condensed milk, porridge with only salt and water, porridge with dried vegetables, porridge with onions, porridge with bananas, porridge with golden-plum juice, and porridge with orange peel. Every degree of consistency from very thin to very thick and every kind of mixture was tried. Up to a point the porridge was always a novelty; but there was invariably an unwanted ingredient, namely reindeer hairs, which could not be got rid of so long as we lived in such close contact with reindeer skins and sleeping-bags. For this reason it was best to eat one's porridge with a spoon and a pair of tweezers.[102]

Lunches and dinners usually consisted of tinned meat and vegetables or ryvita with sardines, repetitive and bland, but special days were celebrated with treats. "We had brought over 40 pounds of whale-meat with us and in its frozen state it kept splendidly," wrote Sorge: "On birthdays and Christmas and Easter a piece of it was sawn off and stewed in a pan with a little water and a lot of butter. It is something like venison and tastes delicious."[103] On Sundays, each of the three men could choose an apple or an orange—frozen solid, of course, but still "we prized this fresh fruit far above any of the other things we had to eat."

Science at Station Eismitte

By establishing a scientific station in the middle of the ice sheet and manning it through an entire year, the Wegener expedition aimed to shed light on four main topics: the meteorological conditions above the central ice sheet, the stratification and physical properties of snow as far down in the ice sheet as possible, the mass balance of the island, and the thickness of the ice in central Greenland. Despite the lack of scientific equipment and supplies at Station Eismitte, and despite the need to modify almost every available instrument so that it would work in polar conditions, the three overwinterers pursued a "full incredible program of observations" in meteorology and glaciology.[104] Seismological studies, however, had to be postponed until the spring since the explosives needed for measuring ice thickness did not make it to the center of the ice sheet before winter set in.

Just after seven o'clock every morning, Georgi cleared the station entrance of freshly fallen snow and climbed outside, notebook in hand, to make his first meteorological observations of the day. "A quarter of an hour later he would come back, often shouting cheerfully, 'New cold record, $-68°F$, not much wind'," wrote Sorge: "Often, however, he would come back covered with snow and complaining of 'Disgusting weather, only $-13°F$, and a 29-mile-an-hour wind'."[105] The low temperatures wreaked havoc on the scientific equipment, freezing the ink of the thermal hygrograph, cracking the wind gauge, and ruining balloon launches. Afternoons were spent repairing instruments and drawing weather charts—a task that Loewe, confined to his sleeping bag, assisted with by performing the necessary calculations in his head. The resulting data—reams of temperature, humidity, and pressure measurements taken daily, and sometimes hourly, at Eismitte through an entire year—laid bare one of Greenland's last secrets: before the Wegener expedition, as the US Geological Survey's senior cartographer François E. Matthes emphasized, "the vast interior of Greenland still was, climatologically as well as geographically, a blank area on our maps."[106] By filling in that blank, the expedition's results attracted attention from meteorological communities in Europe and North America. "It was of the greatest interest to receive complete series of meteorological and climatological records from the station, the only one established for any length of time in the central parts of a continental inland ice," wrote Ahlmann—data that, he emphasized, was collected "under extremely trying conditions, which caused Loewe great suffering."[107]

The strategic importance of this information became clear with the onset of World War II. During the war, Greenland's position between Europe and North America made the island a critical link for meteorological networks, and understanding the year-round weather conditions above the island was essential for the safety of aircraft and convoys crossing the high North Atlantic.[108] And with the war over, Matthes continued, the nature of weather conditions and atmospheric phenomena over central Greenland took on significance for civilian aviation. "The ice sheet of Greenland, the second largest existing on the Earth, lies athwart several of the great air routes of the future," wrote Matthes in 1946, and "factual data concerning the weather conditions with which pilots will have to contend at different times of the year therefore are of vital importance."[109] Even with a US military presence of more than 5,000 men in Greenland during World War II, including more than a dozen US meteorological stations, the Wegener expedition's data still provided the most complete meteorological understanding of the central ice sheet well into the late 1940s.

While Georgi's attention was devoted to the meteorological program, Sorge applied himself to glaciological research. In what Swiss-American glaciologist Henri Bader later called "painstaking work...under the most adverse conditions imaginable," Sorge dug a well 16 meters down into the ice sheet in order to investigate the physical properties of snow layers as deep down as possible.[110] "We had no rope to pull up the loosened ice, nor even a rope-ladder for going down and coming up, and, finally, we had no light," wrote Sorge: "Under these conditions there was great risk of an accident if one fell. For this reason the shaft was made not vertical, but in the form of a flight of steps leading down at an angle of 40° to 60° to a depth of 35 feet, only the last 15 feet being vertical."[111] Just wide enough in diameter to accommodate a man, the well was a deeply uncomfortable place to work. Sorge's days were spent maneuvering up and down the slippery stairs, balancing a small lamp while checking on dozens of thermometers and pressure recorders, his face protected by a black balaclava. By the end of the overwinter, the footholds were smooth and rounded with wear—so much so that Georgi, lacking Sorge's intimate knowledge of their placement and contours, dreaded descending into the shaft.

From the depths of his well, Sorge undertook the first systematic study of snow layers in central Greenland, measuring the temperature, density, and stratification of the snow down the length of the well. To his surprise, it was impossible to see the stratification of summer and winter snow layers as he had been able to do in Alpine glaciers. In order to identify seasonal snow layers, Sorge measured the specific gravity down the well by carving out snow blocks, measuring them, and then weighing them with a Roman steelyard. "The result of [these] measurements was successful," he wrote: "the change of summer and winter layers was ascertained by the changing densities."[112]

By distinguishing the denser winter snow from interweaving lighter summer layers, Sorge's careful measurements demonstrated the feasibility of accurately identifying accumulation cycles in the ice sheet. At the very bottom of his well, Sorge stood on snow which had fallen 20 years previously, in 1911, before Europe had been enveloped by World War I. To travel down in the ice sheet is to travel back in time, and Sorge's work at Station Eismitte provided a first map to guide that journey. Ultimately, Sorge's work showed that, on polar glaciers with no summer melt, the density of snow as a function of depth changes with time only if the climate changes. Described by Henri Bader as "the most fruitful law of polar glaciology," this observation provided a simple method of determining past temperature variations in places and for times with no available direct meteorological observations.[113] In effect, as Bader continued, Sorge's work exposed Greenland's snow layers as a "treasure trove for the scientist," inspiring a long tradition of ice core research at Eismitte and beyond.

Together with meteorology and glaciology, seismic research formed the third branch of Wegener's planned scientific program. After all, the desire to establish German priority over seismological work in the polar world had been a driving force in gaining support and funding for the expedition. Lacking supplies, the overwinterers at Station Eismitte were unable to perform seismological studies until a relief party arrived in the spring of 1931, bearing hundreds of kilograms of explosives. In Greenland, Sorge and geophysicist Bernhard Brockamp led the German effort to measure the thickness of the ice sheet.[114] With his seismological experience at Askania Werke, a precision instrument maker in Berlin, and on Austria's Pasterze Glacier, Brockamp hoped to describe the structure of central Greenland's ice sheet for the first time by setting off explosives and measuring the paths of the resulting waves through the ice. Snow blasted hundreds of meters into the air as his team set off dozens of explosions on the ice sheet, but malfunctioning equipment made for unsatisfactory results. "There is not a single record from Eismitte in which the arrival of the reflected wave is convincingly clear," commented the palpably disappointed British surveyor Michael Spender and Danish archaeologist Therkel Mathiassen in 1934.[115] Still, Sorge was confident enough to paint a broad picture of the ice sheet's shape, telling the Royal Geographic Society in an afternoon lecture in December 1932 that "the thickness of the ice at Eismitte amounts to about 2,000 meters, but there is a possibility of its being anything between 1,800 and 2,700 meters."[116] As it turned out, Sorge's estimate was low: nearly 20 years later, a French team measured the ice thickness at Eismitte to be just over 3,000 meters.

More importantly at a time when Greenland's basic shape was still a mystery, though, Sorge told the attentive crowd that seismological measurements taken at regular intervals on a path from Eismitte to the west coast showed that "one might compare Greenland with a plate filled with ice," with a thin rim and thicker middle—a description that would later be proved correct.

Eismitte Enters the Scientific Imagination

Only decades before the Wegener expedition to Eismitte, the question of ice-free valleys at the center of Greenland was still unresolved. By the 1920s, Greenland's coasts had been mapped, but the central ice sheet was still largely unknown. With their overwinter at Eismitte, Georgi, Sorge and Loewe filled in one of the last major scientific blanks on Greenland's map: with their data, diligently recorded through months of cold and darkness, the meteorological, glaciological, and seismic contours of the central ice sheet took shape. The scientific results of the expedition, compiled and published in seven volumes under Kurt Wegener's watchful eye between 1933 and 1940, met with acclaim from the scientific community—and raised new questions about the shape and depth of the island's substratum, the effects of ice sheet meteorological conditions on the North Atlantic, the dynamics of the ice sheet, and the potential of buried snow layers to unveil information about the past. A thought experiment performed on the basis of the expedition's ice thickness measurements, Danish geographer Einar Storgaard wrote in 1932, showed that "if all of Greenland's ice melted, the world's seas would rise by eight meters, and large tracts of the earth's low-lying areas would be flooded."[117] Denmark, he continued, would be devastated: "Amager and Saltholm would disappear completely, Laaland would be reduced to a small island, large tracts of west and north Jutland would be under sea, and in central Copenhagen only the neighborhood around the Church of Our Lady and the University would remain as an insignificant island." Further investigations at the center of Greenland's ice sheet, he suggested, were needed to fully understand these implications. Eismitte had, indeed, entered the scientific imagination.

Before leaving Eismitte for the last time, Sorge carefully erected a seven-meter high wooden tower to ensure that the German station could be found again even after years of snow accumulation. The tower, Sorge told the Royal Geographical Society in late 1932, was "necessary of course to rediscover the position of Eismitte"—a task which he anticipated would happen within the decade.[118] At that point, just over a year after returning from Greenland—a year marked by widespread acclaim for the expedition's scientific achievements and intense mourning over Wegener's death—Sorge could not have predicted that it would be nearly two decades before Eismitte was revisited. As Europe was again plunged into war, polar work was disrupted and the last marker of the German expedition, Sorge's wooden tower, was slowly engulfed by the snow.

Chapter 2

Taming the Ice Sheet

France's Polar Nestor: Paul-Emile Victor

Newly liberated after World War II, with its infrastructure destroyed, bread rationing still in effect, its scientific establishment in tatters, and facing colonial crises in Indochina and Algeria, France was hardly in a position to mount an expensive mission to the Arctic. But as polar science came back to life after the disruptions of the war, French anthropologist and explorer Paul-Emile Victor set his sights on Eismitte. For Victor, who had spent nearly two years living and trekking in Greenland in the interwar era, the very center of that island represented at once one of the last great Arctic challenges, the ideal culmination of his prewar adventures, and a focal point for broader polar ambitions, which included securing France's Antarctic claim.

Born in Geneva on July 28, 1907, Victor grew up in the mountainous Jura region of France where his father was a small-scale industrialist specializing in pipes and pens. As a boy, he immersed himself in the worlds of Jules Verne, Jack London, and Rudyard Kipling, reading their works over and over and dreaming of great adventures to the ends of the earth. After completing secondary school, or *lycée*, Victor entered l'Ecole Centrale de Lyon, one of the French *grandes écoles* for engineering, which at the time—the 1920s—had a good reputation for educating the sons of industrialists. After three years, however, Victor left Lyon without a diploma, electing instead to join the *Ecole Nationale de la Navigation Maritime* (National Maritime Navigation School) in Marseille. Between 1928 and 1930, Victor attended the Marseille school as a student officer, and subsequently fulfilled his military service as a midshipman in *La Royale*, the French Navy. He quickly grew tired of the mariners' world, though, and in 1933 moved to Paris to study at the Musée du Trocadero (renamed the Musée de l'Homme in 1938). Under the tutorship of French anthropologist Marcel Mauss, Victor was introduced to a style of anthropology that valued prolonged immersion in native societies and rejected geographical determinism.[1] Mauss had spent much of the early twentieth century amongst the native peoples of Alaska and northern Canada, and cultivated Victor's interest in the polar world.[2]

Victor's exposure to polar anthropology in the comfortable confines of the Musée du Trocadero would bring him to Greenland within a year of his arrival in Paris, and ultimately define the rest of his life. Together with three companions (anthropologist and Musée du Trocadero graduate Robert Gessain, Swiss-born geologist Michel Perez, and cinematographer Fred Matter-Steveniers) and armed with an 8,000 franc grant (USD $9,000 in 2012 dollars), Victor spent the winter of 1934–1935 at Angmagssalik (now Ammassalik), a community on a small island off Greenland's east coast. Over the winter, he applied his mentor Mauss' methods to undertake an ethnographic study of the Angmagssalik Inuit. When he returned to France in 1935, Victor recounted his adventures in two popular books, *Boréal* and *Banquise*, whose success brought him into the public eye.[3] It was this ability to make a name for himself that set Victor on a course to build a legacy as "one of the figures of adventure of the 20th century who illuminated the changing culture of a living native people."[4]

It wasn't long before Victor would return to Greenland: soon after arriving back in France, "the fever of adventure took us again" and, in 1936, Victor—along with old companions Robert Gessain and Michel Perez, as well as the dean of Danish polar exploration, Count Eigil Knuth—succeeded in crossing the ice sheet from west to east.[5] The small team departed from Christianshaab (now Qasigiannguit) in May with a team of 32 dogs and arrived at Angmagssalik two months later, having traversed 800 kilometers of ice in 49 days. The expedition was far from smooth: as the volatile ice sheet weather reared its head, the men had to abandon all nonessential supplies, including the collapsible boat intended for use on the eastern coast, and shoot some of the dogs. "Three weeks of poor conditions," read a telegram sent to Paris on July 15, 1936 announcing the expedition's arrival in Angmagssalik: "we were obliged to lighten the sleds and part with some of our belongings and rations. Numerous dogs are dead. We have arrived tired but enchanted."[6] Victor's deep frustration with this expedition, which he later called "the end of the dogsled and foot era [of exploration]," would drive him to return to the ice sheet after World War II.[7]

Having arrived back in Angmagssalik, Victor did not return immediately to France, but rather spent the next 14 months in the eastern

Greenland community. Living, in his words, "in an isolated Eskimo house like a man of the family," Victor continued his ethnographic studies and improved his knowledge of the local Tunumiusut language.[8] Upon return to France, he gained further acclaim in the public sphere through popular publications and speeches with enticing titles such as *Je suis un Esquimau* (I am an Esquimo) and *Impressions du Bout du Monde* (Impressions from the End of the Earth), as well as the publication of his expedition diaries.[9] Victor's confident and warm public personality would continue to serve him through his career, especially in his later efforts to secure funding for his Eismitte expedition. Within two years of his return to France in 1937, however, the opening of World War II forced Victor back into French service and put a temporary halt on his ethnographic and exploratory adventures.

At the beginning of the war, Victor was stationed in Stockholm as an adjoint to the French naval attaché to the Scandinavian countries. He spent several months in Finland after the outbreak of the Winter War between Finland and the Soviet Union in late 1939, spying on boat movement in and out of ports. After the Fall of France in 1940, Victor was released from duty by a government commission giving him permission to pursue two of his passions: ethnographic research and youth movements, an interest that stemmed from his long involvement with Scouting. Victor immediately left Europe for brief stays in Morocco, where he assisted with the reorganization of the Moroccan Scout movement, and occupied Martinique, where he conducted ethnographic studies, before arriving in the United States in July 1941.

Unhappy with his inaction in face of the war, but also reluctant to join the Gaullists in the wake of the Allied invasion of Vichy French-controlled Syria and Lebanon, Victor applied for US citizenship and enlisted with the US Army Air Forces in July 1942. While he dreamed of returning to Greenland as part of a search and rescue team, Victor did not make it back to that island during the war. Recognizing his polar experience, the Army Air Forces assigned Victor to the Arctic section of the Arctic, Desert and Tropic Information Center (ADTIC). Created early in the war, ADTIC was responsible for studying the medical, physiological, and biological challenges faced by military personnel stationed in hostile natural environments. Victor gradually increased his responsibilities: first writing technical manuals for polar survival and travel, then running cold environment training camps in the Colorado Rockies, and finally commanding an aviation search and rescue center out of Nome, Alaska. His experience with the US military opened his eyes to new ways of interacting with polar environments and trained him in practical skills from parachuting to cold regions flying to mechanized Arctic overland transport—all of which he would put to use in his postwar expeditions to Greenland and Antarctica. "Here I became familiar with parachuting techniques and the performance of aircraft and weasels [motorized vehicles] under northern conditions," wrote Victor soon after the war: "I soon realized that by the use of an aircraft with mechanized vehicles it would now be possible to transport the heavy equipment necessary for the work we had contemplated in Greenland."[10]

Following his World War II experiences, Victor was eager to return to Greenland—and to bring modern techniques and technologies to the ice sheet. His envisioned a new age of polar exploration, one rooted in wartime technologies and focused on scientific investigation. His US military experience had convinced him that there should no longer be any significant technological barriers to polar expeditions. It was with this vision that Victor founded Expéditions Polaires Françaises (EPF) as an organization to conduct and champion scientifically oriented French polar expeditions. "There was no question of undertaking such a program, with such aims, with dog sleds and walking," wrote EPF physician Gérard Taylor: the new French organization was to represent the conquest of the ice sheet by modern technologies.[11] It was precisely this use of technology that would, Taylor explained, enable polar expeditions to break away from the older era in which efforts were necessarily focused on survival and arise in a new fashion centered on scientific investigation. "The time of polar expeditions with purely sportive and adventurous goals is passed," he proclaimed boldly: "in our era, the objective has become essentially scientific." These ideas would form the core of EPF's narrative for years to come.

Technologically intensive expeditions, however, had the drawback of being expensive—very expensive. As envisioned by Victor, EPF would not be able to support itself with private funding, but would have to secure extensive government support. To appeal to France's government of the day, Victor painted his nascent organization as essential for his country's prestige and prominence in the polar regions and, specifically, for maintaining French possession and control of its Antarctic claim, Terre Adélie.[12] "The large number of nations interested in the Arctic and the Antarctic proves that these are new regions of the globe now conquered by man who, henceforth, will develop important activities there," wrote Victor in 1947: "Since the war, the principal nations of the world have recognized the strategic, economic, technical and scientific importance of the polar regions."[13] The list of countries active in the polar regions, he continued, included the United States, Canada, Great Britain, the Soviet Union, Denmark, Norway, Sweden, Australia, New Zealand, Chile, and Argentina—but not France. The newfound importance of the polar regions, he argued, justified and even required France to assert itself so that the country did not lose out on the world stage. A French polar expedition, Victor concluded, would improve the country's image on the international scene, enable France to claim a leading role in future international polar endeavors, and "elevate France as a great nation by its contributions to common scientific, economic and aviation problems."

The national interest card was strongest in the context of Terre Adélie, a 6° arc of the Antarctic continent stretching from Pourquoi Pas Point, opposite the Australian city of Melbourne, to the South Pole. French claims to Terre Adélie dated from 1840, when French explorer and naval officer Jules Sébastian César Dumont d'Urville hoisted the *tricouleur* on the land. Dumont d'Urville named the territory after his wife, Adélie Pépin, whom he hadn't seen since setting sail in the *Astrolabe* four years previously.[14] By the late 1940s, Victor lamented, no Frenchman had returned to the territory since its discovery over a century earlier. "We must not lose this land for which French possession is already deeply contested," wrote Victor in 1947.[15] Given the riches of Antarctica and the encroaching interests of other nations, he argued, the need to send a French expedition to secure the country's Antarctic interests was urgent.[16] In a letter to the Ministry of National Economy, written in the same year, Victor reemphasized his argument: before France lay the opportunity, he wrote, to take a leading place in polar science—a position that would, in turn, bring economic and technological benefits, especially with the increasing importance of the Arctic for aviation and the probable advent of regular flights over the pole.[17] Victor's arguments found a friendly reception in government circles, as the National Assembly was already taking steps to establish a Parisian office to oversee France's southern interests.[18] In this context, Greenland was portrayed as a necessary stepping-stone: a polar region close to home and relatively easy to access, and a testing ground for the technologies and logistics which EPF would eventually bring to Antarctica.

Expéditions Polaires Françaises

Expéditions Polaires Françaises was formally created on July 2, 1947, when the President of France, Vincent Auriol, granted Victor's organization his high patronage. With his scientific program approved by two of France's leading scientific institutions, the *Bureau des Longitudes* and the *Académie des Sciences*, Victor was officially charged with studying Greenland's ice sheet and its influences on the northern hemisphere. Along with the President's patronage came state funding. The Ministry of National Education and the *Centre National de la Recherche Scientifique* (National Center for Scientific Research) contributed 40 million francs (USD $1.2 million in 2012 dollars) towards salaries for personnel and the purchase of scientific equipment for Victor's planned Greenland expedition alone.[19] The French Armée de l'Air, too, contributed by providing aircraft and flight crews. The Armée de l'Air's support came in the wake of a lobbying campaign in which Victor, ever the enthusiastic entrepreneur pressing his projects, argued that EPF represented an unprecedented opportunity for the French military to gain experience in cold regions flying and Arctic air operations.[20] Dangling his connections with the United States as a further carrot, Victor portrayed EPF as the Armée de l'Air's ticket d'entrée for observing and interacting with state-of-the-art US aviation technologies on the US bases in Greenland.

Similar lobbying techniques enabled Victor to raise additional funds from the private sector. Victor was a talented self-publicist, convincing companies to support his expeditions and taking to the media to win public approval. Food giant Nestlé supplied EPF with Nescafé and Nescao to drink at Eismitte—"strong comforts in difficult moments," as a company advertisement showcasing its relationship with EPF read—and other companies provided supplies ranging from canned meat to tent cloth to chronometers to cameras.[21] Photographs, films, and books were also sold to a receptive public. This multiplicity of funding sources underlines EPF's dual identity: While officially a private organization, EPF's need for government patronage meant that the organization was not autonomous from public interests.

In the European economic climate of the late 1940s and early 1950s, when postwar reconstruction took precedence, the funding given by the French government to EPF stands out. Following the liberation of France, the provisional government's priorities were to rebuild the country's roads, ports, and cities that had suffered so badly through five years of occupation and bombing, and from German sabotage in the closing days of the war.[22] Science did not factor in these priorities: left out of national economic planning for more than a decade after the war, French science was fragmented, directionless, and short of resources.[23] In this unfavorable climate, it was Victor's appeal to France's territorial interests in Antarctica that convinced the government to invest in EPF. "In the context of France's situation three years after the war, this government decision was tremendously courageous," wrote historian Nicolas Skrotzky: "the country was without means, deprived of laboratories and research teams, and faced with enormous problems of daily life to resolve."[24] Writing from his post as the Danish government's scientific counselor for Greenland in 1951, Lauge Koch underlined the importance of Victor's funding situation to the exploration of Greenland: "no other man and no other nation is capable of completing the work that you have begun," he wrote, stressing that neither his home country of Denmark nor other European governments were likely to provide similar funding at that time.[25]

Having received approval and funding, Victor inaugurated EPF with a series of expeditions to Greenland—"the land of my dreams"—between 1948 and 1953.[26] In doing so, he was following in the footsteps of a number of nineteenth- and early twentieth-century French polar explorers, including Lieutenant de Veau Blosseville, the first Frenchman to set foot in Greenland, who in 1833 disappeared together with his ship along a portion of the coast that today bears his name; Dumont d'Urville of the 1840 Terre Adélie expedition; the Duc d'Orléans, who undertook oceanographic studies in the Greenland Sea in 1905; and Jean Charcot, who overwintered in Antarctica in 1909 and subsequently explored the Greenland Sea and the Greenland-Iceland strait in the late 1920s and early 1930s with his boat the *Pourquoi Pas*, as well as the French team which participated in the International Polar Year of 1932–1933.[27]

Expéditions Polaires Françaises in Greenland: The Eismitte Expedition

Central to the first EPF expedition to Greenland was the establishment of an overwinter station on the ice sheet at Eismitte. Named *Station Centrale*, the French station was a direct reinvention of the Wegener expedition's Station Eismitte. The location was chosen both to pay homage to the German expedition, held in high esteem by the French team, and to extend the scientific and meteorological observations made on the German overwinter two decades earlier.[28] "These two men did not sacrifice anything of their meteorological and glaciological work, which they pursued without failure, with clock-like precision," exclaimed EPF meteorologist and assistant leader Michel Bouché in admiration of Johannes Georgi and Ernst Sorge.[29] Like others on the EPF team, Bouché knew the Wegener story by heart and dreamt of walking in the footsteps of that famous expedition and building on its work. Scheduled to begin in 1948 and run for five years, the EPF expedition to Greenland took as its primary goal to study "this immense ice wasteland, its anatomy, its physiology, or if you prefer, its life and its influence."[30]

EPF's work in Greenland began in the summer of 1948 with a preparatory campaign on the west coast to put in place the logistical apparatus necessary for a planned four years of scientific investigation. Station Centrale itself was established the following summer, and the two overwinters at Eismitte—the highlights of the expedition—took place in 1949–1950 and 1950–1951. French scientists left the central ice sheet station for the last time in August 1951, leaving it to be buried by accumulating snow. During those two years, supplementary traveling expeditions covered more than 8,000 kilometers on the ice sheet, and the next two summers (1952 and 1953) saw reduced summer campaigns in Greenland.

During the summer preparatory campaign of 1948, a team of 26 men traveled by boat to iceberg-studded Disko Bay, nestled at the 70th parallel on Greenland's western coast. Disembarking at Quervain Havn, the men—shirtless from exertion—spent the summer and autumn hauling supplies up onto the ice sheet and beginning to move them inland. With its jagged mountains and unstable ice, the marginal zone of the ice sheet was challenging and risky to cross. The construction of a cable car line across the mountainous ribbon at the edge of the ice enabled the men to move loads over the rocky cliffs in under ten minutes—a feat that would have been impossible by pony, sledge, or even motorized vehicle.[31] Even so, wrote EPF logistician Robert Pommier, accessing the ice sheet required "a festival of patience."[32] All members of the preparatory campaign, regardless of their designated role, were transformed into longshoremen and haulers, and hard manual labor filled the long summer days.

On July 1 of the next summer, the first convoy left the edge of the ice sheet and headed inland towards Eismitte. With 22 people and eight tons of supplies (including two portable laboratories) spread over five motorized tractors, this convoy marked the beginning of a nearly two-month long effort to bring supplies for the French overwinter to the center of the island. Between the crevassed marginal zone and inland blizzards, the voyages were anything but smooth: melt zones halted convoys for days at a time, ferocious winds threw sleds upside down, crevasse bridges fell apart under the weight of the motorized vehicles, and broken chains had to be repaired with bare hands in freezing temperatures.[33] At top speed, the convoy could move across the ice at 40 kilometers an hour, but when the weather was recalcitrant operations slowed considerably: "All those who have known the frustration of starting an automobile on a cold morning can easily imagine the anxiety and pain of the officer who, each morning, faces the problem of starting 20 completely frozen vehicles in temperatures of -40°C and below," complained an anonymous frustrated member of the expedition.[34] Despite these difficulties, all of the materials necessary for building an overwinter station at Eismitte were in place by the end of August: fuel, scientific instruments including a delicate and expensive seismograph, and heavy building materials had been transported 400 kilometers to the center of Greenland.

For the overland travel, Victor chose American weasels, giant amphibious tractors developed for hostile terrain during World War II and described by the US Marine Corps Gazette as "one of the most versatile weapons thus far introduced to war."[35] Over 15,000 weasels were built during World War II and used by troops to travel over snow, ice, and mangrove swamps in theatres ranging from Alaska to Europe to the Pacific. After the war, the four-seat vehicles found homes across the Arctic and Antarctic, transporting scientists and military personnel over the ice and snow.[36] Victor had plenty of experience with weasels from his days in the US Army Air Forces, and the vehicles were easily available through US Army surplus. Still, Victor cautioned, "The unknown opened before us in all its domains: such an enterprise had never before been successfully attempted, and our mechanized methods still needed to be proven."[37]

Upon arriving at Eismitte in the summer of 1949, the French team members were the first men to set eyes on the location of the Wegener expedition's 1930–1931 overwinter camp in nearly two decades. "But what a nerve-racking universe surrounded us!" they exclaimed:

> We were at 3,000 meters of altitude and our fatigue was immediate, our shortness of breath at the slightest exertion adding to the proof of the altimeter readings. But what a strange "summit" it is: everything is absolutely flat around us, as far as we can see. When we leave our tent and turn towards the east, it is 500 kilometers of uniform ice desert separating us from the Greenland Sea. When we turn towards the west, it is 500 identical kilometers to the Baffin Sea from which we came. Towards the south? 1,200 kilometers, similarly uniform. Towards the north? 1,200 kilometers, without an undulation. [It is] the edge of the world…the end of the world. But not even, because this is not the world that man knows.[38]

Station Eismitte itself had been completely buried by snow, and no sign remained of the cave where Sorge, Georgi, and Loewe had spent so many months, or even of the wooden tower erected at the end of the German expedition. But the excitement of the French team to be at Eismitte was palatable: the summer days were "lived in haste, swallowed one after another under the pressure of the work to be done, merged together in memory."[39] The station built that summer was an under-snow installation, completely buried in the ice sheet to give protection from Greenland's wind, cold, and ferocious winter storms. Station Centrale boasted more than 20 rooms including living and sleeping areas, medical facilities, a toilet, garbage storage, a photographic darkroom, radio equipment rooms, a machine room, storage space for fuel, food and hydrogen, and scientific laboratories designed for glaciology, seismology and meteorology, all prefabricated in France from flame resistant panels. In the kitchen, a large dining table covered with a checkered tablecloth and littered with ashtrays stood alongside a two-burner gas stove. A separate hotplate meant that men on rotating shifts could always enjoy a hot meal. With arched roofs made from hand-cut snow blocks and a main snow-walled corridor nicknamed the *métro*, the French station juxtaposed the modern and the raw Arctic. A petrol stove, running continuously but weakly, kept the interior temperature between 0° and 15°C, chilly but tolerable, even as the temperature outside plunged below −60°C.[40] The under-snow scientific laboratories, powered by a generator, were augmented by an aboveground meteorological station and a tower for launching radiosonde balloons, both connected to the station's entrance by a rope to ensure safe passage during the 24-hour darkness of the winter months.[41] In comparison with the Wegener expedition's snow cave, the French station was a veritable paradise under the ice.

Not all of the materials needed to build and supply Station Centrale came in overland. Of the 100 tons of materials, 30 tons came by weasel and the remaining 70 tons were delivered from the sky. The French team saw air support both as a modern method of bypassing the difficulties and dangers of traveling overland on the ice sheet and as a technologically rich symbol to distinguish their expedition from those that came before. Air support, emphasized Charles Gaston Rouillon, the assistant director of the expedition, was the key to the successful installation of a large-scale overwinter station in the middle of the ice sheet.[42] This vision of the expedition as a technologically advanced endeavor was reinforced by journalists from *Agence France Presse* and *Radiodiffusion Française* who, invited to join the air support missions, reported in print and on the radio about a "sensational and exclusive" story in which adventurous aviators employed the latest technologies to sustain the men on the ice sheet.[43]

The air support for EPF was planned by the *Troupes Françaises Aeroportées* (French Airborne Troops) and conducted under the calm eye of former submariner and Air France pilot Roger Loubry. Owing to the dangers of landing on and taking off from the ice sheet, Loubry and his pilots kept their planes airborne at all times over the ice sheet, dropping supplies out as needed. The entire system was designed from scratch and tested in France and in Iceland before being deployed in Greenland: packaging combustible and breakable materials, loading the aircraft, setting up drop zones on the ice sheet, and preventing the dropped supplies from impaling themselves deeply in the snow. Over the summer of 1949, French aircraft based in Iceland flew 13 missions to Eismitte, gradually perfecting a system that would keep the station supplied through the winters with 25,000 liters of fuel, 9 tons of rations, 90 bottles of hydrogen gas, and 20 tons of diverse materials including chemicals, explosives for seismic experiments, and Christmas packages.[44] It was thanks to this system, wrote seismologist Jean-Jacques Holtzscherer and glaciologist Albert Bauer, that "the EPF were able to pursue work on a much vaster scale then the work undertaken 20 years before by the pioneers of Wegener's expedition."[45]

Scientific Overwinters at Eismitte

On August 24, 1949, the summer team departed Station Centrale, leaving behind eight overwinterers led by French bobsleigh champion Robert Guillard. Made up of glaciologists, meteorologists, physicians, radio operators, and mechanics, the overwinter teams were handpicked by Victor to realize his long-held dream of taming the ice sheet. During the two winters, the men were alone at the center of the ice sheet:

> The white desert, infinite, uniform, crushing, opened before us. No place in it was reserved for man. The Sahara has its oasis; the Himalayas have their green valleys; and the Matto Grosso its swarming life. But the inland ice, no matter which face it shows man, cripples, kills. Man is not made to support temperatures of $-70°$, to fight against winds of 180 kilometers an hour, to cope with the blinding reflection of sunlight in a mirror without an undulation, without a wrinkle, without a shadow in the snow, to resist the monotony of long months of solitude at the heart of an immutable infinity, inhuman, not to his scale and which can offer him as a distraction only the movement of the thermometer.[46]

The biggest challenges facing the overwinterers were the relentless cold and the crushing monotony of the days, each one identical and regulated by a strict schedule. The initial excitement quickly wore off as bottles of ink, wrapped in woolen socks, froze into useless blocks and the bare hands and fingers necessary for manipulating a pencil grew too cold to write. Facing temperatures of $-30°C$ in the corridors and an average of $10°C$ in the living spaces and laboratories, the chill was constant and inescapable.[47] Temporary relief came in the evenings, after dinner, when music broke the silent and still atmosphere of the ice sheet and radios brought news from the men's families in France. The team played their favorite albums—Rachmaninoff's concertos and Paul Robeson's spirituals—and smoked cigarettes, "which we consume abusively," a brief respite from a demanding scientific schedule.[48]

The French team tried to battle the cold, the repetition, and the long, dark days with pleasures of the palate: by working in collaboration with the French Institute of Food Hygiene and Stewardship to develop new rations and reduce reliance on monotonous tinned foods, wrote expedition radioman Mario Marret, EPF aimed to supply the calories needed to fight against the cold and the vitamins necessary to maintain good health while also providing a variety of tastes and textures to prevent gastronomical boredom from afflicting the men. "The plate of the explorer," he boasted, "contains the ideal alimentation, and ideal quantity"—but, in reality, even careful planning could not prevent food from becoming a source of contention among the overwintering men.[49]

At Station Centrale, the regular daily diet consisted of typical ration foods, lightweight, easy to store, and long lasting: dried meat, rice, potato powder, milk powder, vitamin drinks, sugar, chocolate, spices, porridge, biscuits, jam, and pasta. To these the French added cheese, butter, copious quantities of cigarettes and hundreds boxes of Nescafé instant coffee. While most of these rations were openly available to team members to consume as they pleased, meat and vegetables were in short supply, as was flour for bread, and kept under figurative lock and key.[50] This diet was the chief cause of disagreements between expedition members: the daily rations were plentiful, but not particularly appetizing. When scientific instruments acted up—not a rare occurrence—the cooking (shared between the men on a rotating basis) "suffered from a failure of imagination and we ate rice—served plain."[51] Nights of bad food deflated morale and sparked arguments. Special foods, carefully stored and allocated separately from the daily rations, went some ways to improving the atmosphere. When rare cans of salmon and sardines, or some of the 20 bottles of wine and rum reserved for celebrations, were brought out, dinners became jovial, unifying affairs. And when an airdrop brought fresh produce—radishes, tomatoes, lettuce, and fruit—the expedition team cooked up a feast, finishing late in the night with slices of melon dipped in melted chocolate.[52]

The EPF expedition to Eismitte was, first and foremost, a scientific endeavor. With a comprehensive scientific program planned out meticulously in advance, the French expedition looked to study the physical characteristics of Greenland's ice sheet and its influence on the northern hemisphere. At Station Centrale, the scientific program included research in glaciology, seismology, meteorology, and geodesy, and on the ice-free coasts geological, geomorphological, entomological, zoological, and botanical studies were added. Of main concern to this chapter are the scientific investigations conducted during the two overwinters at Station Centrale in 1949–1950 (eight men) and 1950–1951 (nine men). "The first grand lines of the scientific program I have established go back to the Wegener expedition," wrote Victor, and, indeed, his program aimed to augment the scientific data obtained by Johannes Georgi, Ernst Sorge, and Fritz Loewe during their 1930–1931 overwinter by ensuring copious and regular data-collection uninterrupted by more pressing demands of survival in a polar environment.[53]

During the first overwinter, the meteorological program was headed by an unlikely character: Michel Bouché was in the process of finishing his studies as a lawyer at the *Institut d'Etudes Politiques de Paris* (Paris Institute of Political Studies, better known as *Sciences Po*) when he heard about the expedition. Tempted by the adventure and armed with a meteorological background from his military service, Bouché volunteered to join Victor's team. At the center of the ice sheet, he ran the meteorological program with military precision: seven times each day, Bouché made observations and transmitted the data and weather reports by radio back to Paris, where meteorological maps of the Eismitte environs were drawn up at the EPF headquarters, a rabbit warren of offices, laboratories, and equipment storage rooms stretching over half a hectare. Using his bare hands to inflate and launch weather balloons, Bouché was able to collect temperature and air pressure measurements from the top of the troposphere.[54] By the end of his overwinter, Bouché had amassed 1,500 pages of observations taken hour after hour, amounting to some 30 kilograms of documentation about the weather at Eismitte.[55]

Since Ernst Sorge's discovery of a link between snow density and annual accumulation cycles in 1930–1931, the idea of digging or drilling deep into polar glaciers had been a priority in the glaciological world.

The disruptions of World War II, however, meant that no glaciological teams pursued drilling research again until the late 1940s. Between 1949 and 1952, three teams inaugurated ice core drillings in three different parts of the world: the British-Norwegian-Swedish Antarctic Expedition drilled a ~100 meter ice core in Queen Maud Laud between 1949 and 1952, the US Juneau Ice Field Research Project extracted a core of similar length from Alaska's Taku Glacier between 1949 and 1950, and the French EPF drilled at Eismitte. This work marked the beginning of an intensive international effort to improve drilling technologies and use ice cores to understand past climates.[56]

As early as 1947, the French were eager to extend Sorge's work by digging further down into the ice sheet and conducting systematic stratigraphic studies of the snow layers.[57] To push further down into the ice, the EPF team developed a hydraulic diamond drill that could be powered by a weasel motor. In August 1950, after 23 days of drilling, the team recovered a 151-meter long ice core at Station Centrale—by far the deepest ice core in the world at that time. The delicate core, barely five centimeters in diameter, was to be analyzed to determine the temperature and density at various depths.[58] While the EPF ice core was successful in terms of length, however, it was of poor quality and proved difficult to analyze: the drilling motion had compressed the core samples to such an extent that measurements were deemed unreliable. The core's immediate scientific value was hence low—a disappointment to a confident team—but the proof of concept offered by such a deep ice core provided a positive stimulus for the continuation of ice core research. It would take several more years for drilling technologies to improve to the point where ice cores could offer detailed insights into past climates.

The EPF glaciological team also took up a challenge issued by Swedish glaciologist Hans W. Ahlmann in his 1948 address to the Scott Polar Research Institute. Polar glaciology, Ahlmann told his audience, needed to move beyond the realm of natural history by systemically collecting exact measurements of physical ice sheet characteristics.[59] The French team took Ahlmann's words to heart, conducting four years of accumulation, ablation, temperature, movement, and ice thickness measurements at Eismitte and along more than 8,000 kilometers of ice sheet traverses. Slowly, a comprehensive picture of the physical properties of Greenland's ice sheet emerged. Further, before departing Station Centrale for the last time in August 1951, the EPF team mounted dozens of markers in the snow to give future groups a means of measuring the movement of the ice sheet and the accumulation of snow at Eismitte—and they also buried a food cache in the expectation that more expeditions would follow. The French hope of instigating a long-term project to track the ice sheet's movement indeed came to fruition, although not in the way the French had initially expected.[60]

In the decades leading up to the EPF expedition to Greenland, several stabs were taken at estimating the ice sheet's thickness and volume, including the Wegener expedition's efforts at Eismitte, Strasbourg geologist and palynologist Georges Dubois' 1931 estimate, and Yale geology professor Richard F. Flint's 1948 theoretical calculation.[61] Varying widely, these early estimates suffered from finicky instruments and a lack of certainty. A chief scientific aim of EPF was to conduct precise and reliable seismological measurements of Greenland's ice thickness in order to resolve the earlier estimates and build a first substratum map of the island. To do so, the French seismological team took hundreds of measurements not only at Eismitte but also over much of the central and southern portions of the island. These measurements showed that the ice sheet is shaped like a convex lens hemmed in on the eastern and western edges by coastal mountain ranges, and underneath by a rocky substratum—the same shape that Ernst Sorge had tentatively proposed 20 years previously. The scope and scale of EPF's work gave their ice sheet volume estimate of 2,750,000 cubic kilometers, which was well above previous estimates, authoritative weight—and, indeed, the French estimate is within a 5 percent error of today's estimate of 2,850,000 cubic kilometers. Were this ice all to melt, Albert Bauer and Jean-Jacques Holtzscherer asserted in their scientific report, the world's seas would rise by an average of seven meters.[62] While Bauer and Holtzscherer did not take this idea any further, their framing of the ice volume calculation in terms of the consequences of polar warming was picked up on in political and public spheres: in a popular 1954 piece titled "The Top of the World," for example, *The Reader's Digest* described the potential sea rise as a threat to "many of the world's greatest seaports," a sentiment echoed a few years later by Danish Prime Minister Viggo Kampmann.[63] This result was a harbinger of an environmental theme that would gain importance in the following years.

Coupled with the Wegener expedition's earlier work, EPF's scientific work transformed Eismitte into the ice sheet location with the most complete record of geophysical observations. For Ahlmann, the French commitment to his vision of glaciological research was affirming: by helping to elevate polar glaciology into a quantitative, data-driven endeavor, he wrote of EPF, "they bring immeasurable value for international science."[64] "Of special interest to me as a geographer and geologist are your epoch-making investigations of the thickness of the inland ice and your investigations of the land surface below the inland ice," wrote Lauge Koch to Victor in November 1951. "Actually your achievements so far may be compared to the mapping of a large unknown land area about which nobody knows whether it is a mountainous country, a low land, or how large parts of it occur below the surface of the sea," he continued: "[this work] will in future be of immense importance for our conception of the effects of an ice age on the land surface in N. Europe and N. America."[65] Koch's comments reflected a broader view that EPF's scientific work opened the door to a complex set of environmental and climatic questions—questions that would, in the following years, drive further expeditions to Eismitte.

Gaining Access: Negotiations with Denmark

Given the nature of Greenland's political situation immediately after World War II, which saw a Danish desire to reassert sovereignty in the wake of a virtual US takeover during the war, foreign access to the island was a delicate topic in Denmark.[66] Victor knew that he would need to approach Danish authorities carefully in order to gain permission for his expedition. While he was initially confident that his contacts from his prewar expeditions to Greenland would open doors for him, Victor became concerned when he received a letter from Ahlmann warning of growing Danish sensibilities. Ahlmann's own plans for a joint Swedish-Danish-Norwegian expedition to Greenland had just been dashed by Denmark's commitment to scientific nationalism, and the Swede was attuned to Danish sensitivities.[67] "We understand here in Sweden the serious geographical position of Greenland and its consequences on the political situation in the Arctic," wrote Ahlmann, warning Victor to avoid "enter[ing] into any such obligations to the American military forces over which Denmark could react."[68] Ahlmann was worried that the Danish authorities, concerned about domestic perception of US strength in Greenland, would react badly to proposed French use of US military facilities on the island. Heeding Ahlmann's words, Victor offered to bring Danish scientists into the fabric of his expedition so as to minimize Danish concerns about foreign nationals on the ice sheet and to distance himself from the long US shadow over the island. Outright collaboration with Danish scientists, however, proved impossible in light of Denmark's postwar priorities, which placed scientific nationalism in Greenland above other interests in the island. "In reality it cannot be done," wrote Danish explorer Eigil Knuth in response to Victor's enquiry: "The conditions after the war have caused that the Danish work in Greenland must have a more national character than ever before. The idea of our enterprise is to make an independent Danish contribution in Greenland to show the world that Denmark knows her responsibilities and is able to assert them. This is the reason that we cannot even incorporate foreign members on our staff and collaboration with another nation is for the moment impossible."[69]

With a joint expedition ruled out, the French engaged in a careful diplomatic dance with Danish authorities to gain access to Eismitte. In the spring of 1947, the French Embassy in Copenhagen began to negotiate permission for Victor's Greenland expedition with the Danish Ministry of Foreign Affairs.[70] Drawing on his prewar contacts, Victor himself wrote several times to Eske Brun, vice director of the Greenland Office, or *Grønlands Styrelse*, even traveling to Copenhagen to discuss the matter personally with Brun.[71] Forceful and direct but softened by an underlying gentle humor, Victor went to great lengths to assure Brun that the expedition would be entirely self-sustained and that Denmark would incur no costs. The French would bring in no alcohol or wine for the Greenlanders, continued Victor in recognition of one of Brun's chief concerns. By emphasizing links between EPF's planned scientific program and Danish research, including Eigil Knuth and Ebbe Munck's concurrent expedition to northern Greenland, Victor also painted his expedition as valuable to Danish scientific interests.

By the autumn of 1947, these diplomatic efforts had borne fruit and Victor's expedition to Greenland was approved by all necessary Danish bodies: the Foreign Ministry, the Greenland Office, and the Commission for Scientific Research in Greenland.[72] The Danes had decided that the French expedition posed a threat neither to Denmark's scientific work in Greenland nor to domestic perception of Danish control over the island. With permission, however, came stipulations: the French were told to transfer 40,000 Danish kroner (USD $80,000 in 2012 dollars) to the Danish government "to guarantee any expenses which will eventually be caused to the Danish state by the presence of the French expedition on Greenlandic territory."[73] To keep a close eye on the French expedition, the Danes also required that "an expert representing the Royal Government will have to be attached to the expedition with authority to control the activities of the expedition."[74] Victor readily agreed to these two conditions, welcoming Jens Jarl, an engineer trained at the *Københavns Tekniske Skole* (Copenhagen Polytechnic School), to the French expedition.

However, the Danish insistence that EPF use only Danish-monitored channels for radio communication—not an unreasonable demand for a country striving to assert sovereignty—struck the French as a step too far. Victor was distinctly unhappy with this demand, arguing to Brun that "this would mean for us a very great delay in our transmissions and a great added expense," and worrying privately that such an arrangement would put scientific data in the hands of the Danes before it could be collated and analyzed in Paris.[75] Following a flurry of letters and a face-to-face meeting with Brun in Copenhagen in early 1949, Victor succeeded in gaining permission to install a French radio station in Greenland to transmit meteorological observations and other information directly to Paris, on the condition that the Danish representative on the expedition, Jens Jarl, approve all telegrams in advance—a solution that proved acceptable to both parties.[76]

While Jarl turned out to be the opposite of a bureaucratic weight around EPF's neck—the engineer made fast friends with the French team and assisted enthusiastically with glaciological and seismological work, as well as with the more menial tasks of the expedition—his story cannot be untangled from the tragedy that struck the expedition in the summer of 1951. While conducting seismological work in southeastern Greenland near Mont Forel on August 4, 1951, Alain Joset, chief of EPF's seismic survey section, and Jens Jarl were traveling in the lead weasel when their vehicle plunged out of sight into a deep crevasse. In a rare stint of poor planning, the rescue materials were contained in the lead weasel, preventing any immediate rescue effort by the men in the following vehicles. With no possibility of rescuing Joset and Jarl themselves, the remaining men radioed for help—a plea that went first to Station Centrale and then on to Reykjavík. When a B-17 US Air Rescue plane arrived bearing an American rescue crew and Paul-Emile Victor himself, it was determined that the weasel was wedged upside down 25 meters down the crevasse and that the men had died on impact, their bodies falling to the bottom of the crevasse 60 meters below the surface.[77] "They probably did not suffer," wrote Victor in a long letter to Jarl's family: "they may not even have been aware what was happening."[78] Repatriating the bodies proved impossible as the weasel was suspended in a precarious position, threatening to slip further down the crevasse at any moment. The French and American rescuers were, however, able to retrieve diaries, scientific notes, and observation books from the fallen weasel. Leaving the men there, Charles Gaston Rouillon wrote later that year, was "the most dignified tomb for their mission and their sacrifice."[79] Before leaving the site, the remaining men erected a bamboo cross and left a note reading "Here lie in the very spot where they fell, at the bottom of this crevasse, during the accomplishment of their scientific mission, on August 4, 1951: Jens Jarl and Alain Joset."

Crafting Narratives

From behind a metal partners desk in his Parisian office overlooking l'Avenue de la Grande-Armée, a room filled with souvenirs of his expeditions—maps of the polar regions, carvings, walrus teeth, and a model airplane among them—Victor used his publicity skills, honed during his prewar expeditions, to craft EPF's image. The French expedition to Eismitte, he announced in 1947, was designed to be "the last spectacular expedition undertaking the last set of incomplete scientific investigations in the whole of the Arctic."[80] "The Expéditions Polaires Françaises, organized and directed by Paul-Emile Victor, has realized a series of scientific expeditions unique in the world," boasted EPF literature in the mid-1950s: "They have developed new technologies, bringing back important results in all scientific domains. They have permitted France to take its place in the first rank of the international polar domain."[81] Highlighting EPF's embrace of modern technologies, Victor showcased his organization as "the first to introduce to scientific research in the polar regions new techniques discovered and developed during the last war, such as the use of motorized tractor convoys, air transport, parachutes, and new materials."[82] The comfortable and well-organized French station at the center of Greenland's ice sheet marked a new chapter in polar exploration, Mario Marret continued: one based not in sacrifice and adventure but in the efficient production of scientific knowledge. With specialized technicians taking care of all logistical concerns, EPF's scientists were free to focus fully on their research. "It wasn't many years ago that the only polar problem consisted of surviving," explained Marret: with the EPF model, "the role of polar technicians is to make the scientists forget about this vital problem."[83] This narrative, carefully crafted, presented a deliberately distorted picture of the polar sphere: much of what Victor and EPF claimed as their own had been introduced to the polar regions in the interwar years (which saw air operations by, among others, Americans, Danes, and Russians, as well as a strong division of labor in Russian polar work), or was simultaneously being introduced by others in the early postwar years (which saw Americans and Canadians using weasels across the Arctic, and an emphasis on dedicated polar technicians in US polar work). Such a narrative, however, was central to EPF's image and success.

All three aspects of the EPF narrative—scientific priority, technological conquest of the polar environment, and France as a world leader—were cultivated and maintained through regular exhibitions and speeches, media interviews and articles, and the sale of books to an interested public. Victor starred at events ranging from a gala evening with films advertised as the *Club of Explorers Gala: Paul-Emile Victor and the War at 40° Below Zero* to an exhibition at the Danish Embassy in Paris titled *Groënland—Le Danemark Arctique* (Greenland—The Danish Arctic), designed to highlight the "modern" nature of Victor's expeditions.[84] Through the 1950s and 1960s, Victor promoted his expeditions in outlets ranging from the patriotic French exploration magazine *Cahiers des Explorateurs* (Explorers' Notebooks) to the popular science magazine *Atomes*, and even by asking the French Embassy in New York to mail out 1,800 copies of a report about the Eismitte expedition to contacts in the United States.[85] In the United States, he also took to the conference circuit, traveling and promoting his work throughout the country for three months in late 1954 and early 1955. Victor's children's book, illustrated with his own pencil drawings of Inuit, kayaks, and the polar north, brought his exploits to French schoolchildren.[86] Short films and documentaries realized by Mario Marret (who, in addition to being EPF's radioman, also moonlighted as an amateur cinematographer) and French producer Fred Orain attracted even broader audiences.[87] Combining footage of ice sheet dangers and triumphs with interviews in Victor's imposing Parisian office, the films depicted EPF's leaders as tough, courageous, and unwilling to be defeated by the ice sheet. A short film about the 1949 expedition to Eismitte even featured at the Cannes Film Festival in 1952, where it won a documentary prize.[88] These narratives were styled to suit their publics, with scientific, political, and technological stories crafted to elevate EPF as a leading polar outfit on the national and international scenes. Always appearing prominently on maps, posters, films, and even the many French territorial stamps issued in honor of the expeditions, Victor's name became synonymous with Expéditions Polaires Françaises. He was lauded in books as a "dispenser of dreams which allow millions of urban dwellers to experience the mesmerizing dizziness of icy places" and sought after as a public speaker.[89]

This carefully manufactured image took hold in both the public and scientific imaginations, and the French polar organization developed a reputation as a technologically advanced, modern exploration outfit. "Technically, Victor's expedition had marked a revolution in expeditions to the ice cap," exclaimed Danish glaciologist Børge

Fristrup: "Drudgery had been replaced by modern comfort: there was even a wine cellar at the winter station!"[90] Embedded journalists painted the harsh conditions faced by the French scientists—"the fatigue, the cold dig into the bones of our faces, tracing lines like crevasses, shattering lips and letting blood escape," wrote *Radiodiffusion Française's* Georges de Caunes—and the technological modernism of the Eismitte expedition as two sides of one coin.[91] Thanks to the success of that expedition and to Victor's ceaseless publicity work, EPF was much admired for the "time-saving business techniques" it applied to polar expeditions.[92] In the following decades, the superpowers both turned to EPF for help with their own polar work, and countries from Japan to Australia to Belgium explicitly modeled their postwar Antarctic research organizations on the French outfit.[93] EPF continued to work and grow, too, despite internal divisions amongst its leaders: in its first decade, the French organization conducted ten expeditions to Greenland and Terre Adélie involving more than 100 participants.[94] The next decade saw another two dozen expeditions to the polar regions, as well as two expeditions to Iceland's Vatnajökull ice dome. By 1967, over 1,000 scientific and technical personnel had taken part in EPF expeditions, traversing more than 250,000 kilometers of ice and snow. With regard to France's territorial interests in Antarctica, which Victor had capitalized on to gain funding for his Eismitte expedition in the late 1940s, EPF installed a permanent base in Terre Adélie in 1956 and the sliver of Antarctica was designated as an official territory.[95] And for Eismitte itself, the systematic scientific data recorded by the French overwinterers and the environmental and climatic questions they raised drew American eyes to the center of the ice sheet, setting the stage for the next expedition to "middle ice."

Chapter 3

The Longest Trek

The Cold War Comes to Greenland[1]

Situated on the shortest great circle route between the industrial centers of the United States and the USSR, Greenland lay on what threatened to be the hottest path of the Cold War. As the US military planned the defense of the North American continent in the face of a new enemy—one believed to possess superior cold-regions warfare capacities—the vast island stood out as a strategically vital region, largely unknown and uncontrolled, hanging in the balance between the superpowers. It is "imperative" that Greenland be given "high priority attention," wrote US Air Force Lt. Col. Emil Beaudry in a top secret 1949 report on high-latitude defense, because the island "now constitutes one of the most vital areas affecting security of the United States."[2] Greenland was, Beaudry continued, "probably more vital to the defense of the United States than any other single polar area." Military control of Greenland "would provide the United States with bases which would enable our Air Force to better intercept a transpolar attack and to retaliate promptly and effectively against any nation or combination of nations capable of waging modern war," he advised his masters in the Air Force: "our military control of the Greenland Ice Cap would offset Russia's general dominance of the Arctic, provide bases for refueling and maintenance of our aircraft, provide much-needed meteorological stations and most important provide bases for the United States Air Force to conduct Arctic research operation and training." The polar regions are "a likely avenue of approach for untold destruction," continued Beaudry, and "unless guarded could well spell doom for the United States as a nation." If one country was able to "completely master" the island of Greenland, he concluded, "she would possess a new weapon that could not be countered or molested."

As the 1950s opened, Beaudry's view gained wide credence among US military and other polar strategists.³ At the Arctic Institute of North America, an influential research institute, think tank and US military subcontractor, it was taken as fundamental that those concerned about North American defense must be prepared for "the Soviets [to] suddenly cross the snow-covered, rugged Arctic with their modern engines of war."⁴ This fear was compounded by a recognition of the historic preeminence of the Soviets in cold regions. US planners were acutely aware that the Soviet capacity for cold environment warfare had caught the Wehrmacht by surprise and did not want to make the same mistake. "We must make maximum use without delay of the immensity of Arctic space to push outposts northward as far forward as possible and capitalize on the resulting manifold military rewards," continued the Arctic Institute of North America—a belief that underpinned two decades of US military and scientific operations and installations in the far north. US General and former Chief of Army Intelligence Arthur G. Trudeau concurred, writing that "the polar arctic, because it lies astride the shortest route between the industrial heart of the North American Continent and the USSR, has taken on a new military significance for the Soviet Union is expending great energy learning how to overcome the natural barriers of the polar regions. To combat such a threat it is necessary for us to investigate the feasibility and techniques of military operations in this area."⁵ At NATO, too, Greenland was considered essential for the defense of the North American continent. The presence of US bases and early warning facilities in Greenland, asserted a NATO military assessment of Soviet bloc strength and capabilities, meant that "Soviet planners could not expect attacking bombers to reach North American strategic nuclear bases in time to prevent the launching of a large-scale retaliatory attack upon the USSR."⁶ "The defenses in Greenland," agreed Canadian General Charles Foulkes at a 1954 NATO military meeting, formed part of "the greatest deterrent to war at the present time [which] is the retaliatory capacity which is in the United States Strategical Air Force."⁷

Greenland's newfound importance to North American continental security was complicated by its harsh geography and climate, as well as by the island's political situation. Greenland is "the seat of some of the most extreme and objectionable weather in the world," wrote Beaudry—something he knew well from his dramatic rescue of the crew of a crashed American C-47 aircraft near Søndre Strømfjord (now Kangerlussuaq Fjord), on Greenland's southwestern coast, in 1948.⁸ With temperatures reaching below -50°C, 24-hour darkness through the winters, and storms which whipped across the barren landscape, Greenland had little to recommend. The US experience during World War II proved that its armed forces were barely capable of dealing with cold weather. In the battle for the island of Attu, for example, 1,148 American soldiers were wounded and another 1,200 incapacitated from cold weather injuries, "a 1:1 ratio that could not have been sustained for longer campaign."[9] Along with other cold weather near-disasters, including the experience of the US Army 10th Mountain Division and post–World War II military exercises in Alaska and upper New York State, the Aleutian Islands campaign convinced the military of the "extreme and almost insurmountable difficulty" of operating in cold regions.[10] Indeed, in a 1948 interview with *The New York Times*, Army Ground Forces Commander General Jacob L. Devers complained that in cold environments "so much of the soldier's time was spent…in a fight for survival that he had little time to combat the enemy."[11] Clearly, Greenland's environment posed difficult challenges to the US military—challenges that would, early in the Cold War, be met with significant scientific resources.

Science and Strategy in the Polar North

In line with the scientific and technological optimism which underpinned US thought through the postwar era, the geophysical sciences were seen as vital for establishing strategic control of the Arctic.[12] "Science," asserted the Arctic Institute of North America in 1957, "will permit our use of Greenland as an Arctic sword and shield— a mighty bastion of deterrent power essential to the NATO concept."[13] "Expanded Arctic *research* is essential so that we may extend our northern military frontier to the Soviet Arctic littoral," the think tank continued.[14] The result was a far-reaching program to build a scientific understanding of Greenland's ice sheet. Through the Cold War, the US military conducted glaciology, seismology, meteorology, geology, and ionospheric research in Greenland, both autonomously and in conjunction with Danish scientists.[15] This science-based polar strategy was spearheaded by a handful of military bodies designed for the purpose. These include, most importantly, the Snow, Ice and Permafrost Research Establishment (SIPRE), founded in 1949 as part of the US Army Corps of Engineers, and the Cold Regions Research and Engineering Laboratory (CRREL), formed in 1961 through the merger of SIPRE and the US Army Arctic Construction and Frost Effects Laboratory (itself dating from 1953). These organizations were mandated to "serve the needs of the National Military Establishment" by conducting basic and applied research in snow, ice, and frozen ground.[16]

Even at a time when the US military invested widely in basic scientific research, the case of SIPRE's approach to glaciology stood out. At the opening of the Cold War, US military personnel considered fundamental knowledge of the properties of snow, ice and permafrost to be "unsystematic and unsatisfactory"—so much so that SIPRE was compelled to start at the very beginning.[17] The country's neglect of polar regions prior to World War II and the military's experience during the war left no doubt of the need for basic research: "the ad hoc approaches to engineering in cold regions by the Army during World War II had failed many times due to the lack of understanding of the complex behavior of snow, ice, and frozen ground," emphasized US glaciologists Carl S. Benson and Charles R. Wilson.[18] The extent of military need for basic glaciological research was "unusual [and] made necessary by the fact that wide areas of knowledge of the fundamental properties of snow and ice have never been established by civilian research," wrote Yale University geologist and military advisor Richard F. Flint in a 1950 memo on snow, ice and permafrost in military operations, underlining the striking lack of scientific knowledge about glaciology in the United States at the time.[19] "The Department of Defense has been obliged to bring basic knowledge of snow and ice up to the level of knowledge of other substances, such as steels and other metals," continued Flint: "Usually military research does not have to reach so far down toward the foundations of science as it has to do in this exceptional case." In this context, it was glaciologists who led the US effort in Greenland: by studying the movement of glacier ice, the structure of snow, and the accumulation, ablation, and melting of Greenland's ice sheet through the changing seasons, and by measuring the temperature, density, stratification, and grain size of snow layers, among other projects, glaciologists provided the foundations on which practical military work relied. Indeed, to Danish glaciologist Børge Fristrup, the US Cold War effort in Greenland represented "the most intensive...glaciological research program that had ever been envisaged."[20]

The military's dissatisfaction with the state of glaciology in the late 1940s and 1950s reflected the academic and professional status of the discipline in the United States. In the early postwar era, the United States boasted few glaciologists and little homegrown glaciological knowledge. "The United States, having acquired the non-contiguous territory of Alaska in the late 19th century," writes historian Fae Korsmo, "was hardly an Arctic nation," and, consequently, US glaciology was a "relatively small and unorganized field."[21] Across the entire country, only 40 academics identified themselves as doing glaciological research, and in only two of these cases did glaciology make up the entirety of that man's research. "The graduate students now engaged in this discipline are not even as numerous as the faculty," bemoaned Ohio State University's Richard P. Goldthwait: numbering under 30, these graduate students made up fewer than 5 percent of all geophysics graduate students in the United States.[22] In the 1950s, no US schools granted degrees in glaciology and only two universities had cold laboratories devoted to the discipline. A glacial geologist with extensive field experience stateside and in the Arctic, Goldthwait had a deep desire to build a strong glaciology community in the United States and felt his country's weakness in the discipline keenly. "There are virtually no experienced senior academic men devoted chiefly to glaciology," he continued: "No American university is fully equipped for glaciological teaching or research [and] research in the United States is inadequate to attack any of the pressing glaciological problems which are being solved increasingly in Sweden, Great Britain, Norway, Denmark, France, and Austria." On this last point, US science administrator Vannevar Bush agreed readily: countries including Switzerland, Sweden, and the USSR, he wrote in 1947, are miles ahead of the US in terms of glaciology and especially the relationship between glaciology and climate.[23] Tellingly, the principal figure of US glaciology in the 1950s and 1960s, Henri Bader, was a Swiss immigrant who arrived in the United States after the war, having completed his glaciological training in Switzerland.

By the early 1950s, US interests in Greenland ranged from building roads at the boundary of the ice sheet to facilitating long-distance, year-round transport of troops and equipment over the island, and from constructing military camps, radar stations, and rocket launching pads on and in the ice to maintaining snow runways for heavy aircraft—anything, in short, related to military operations on and control of the ice sheet. The first step in these endeavors was to understand the basic properties of snow, ice and permafrost, and it is on this basis that the US Army Corps of Engineers funded millions of dollars to SIPRE and other military research and development bodies. It is difficult—and probably fruitless—to draw sharp lines of separation between "basic" and "applied" glaciological research in Greenland: from a ground-up perspective, projects of all types were worked on by civilian and military researchers, and, from a top-down perspective, military strategists conceived of a continuum of research activities, with more basic endeavors necessary for informing and enabling more applied endeavors. Nonetheless, it is important to emphasize that the lack of US knowledge about cold regions in the late 1940s and early 1950s meant that SIPRE had to devote significant resources along the entirety of this continuum. The US glaciological research conducted in Greenland during the early Cold War can be divided into two main types: stationary research projects and roving, or traveling, research projects. Stationary projects consisted of multiyear (and often year-round) research conducted at specific ice edge or ice sheet locations, including Thule Air Base, Camp Tuto, Camp Century, and smaller research outposts such as Camp Fist Clench and Site 2.[24] At these locations, teams of scientists and research personnel—typically ranging from five to 25 men—undertook research programs in five main areas: fundamental snow, ice and permafrost studies (e.g., glacier movement, physical properties of snow, ice and permafrost, and stratification studies), overland transport (e.g., snow and ice trafficability, pathfinding and trail marking, crevasse detection, and buried railways), air transport (snow and ice runway construction), construction (e.g., ice tunnels, snow stabilization and strength, pile and load studies, and ice deformation over time), and ice core drilling (including deep core drilling to the bottom of the ice sheet).

From the opening days of the Cold War, glaciological research at stationary locations was supplemented by roving expeditions. In a series of military–scientific operations far removed from the comfortable confines of US bases and camps, personnel from SIPRE and other military bodies conducted ice sheet expeditions in strategic areas of Greenland. The first of these expeditions, Project Snowman, which took place in 1947 near Søndre Strømfjord, studied the surface of the ice sheet and investigated the possibility of landing heavy aircraft directly on the ice. This was followed by a variety of expeditions with suitably Arctic names, from the 1953 Operation Icecap, which studied the physical characteristics of the ice sheet and the marginal zone in the Thule and Inglefield Land regions, to Project Mint Julep, which saw snow stratification and ablation studies take place from a temporary campsite in southwest Greenland, to Operation King Dog, a 1958 reconnaissance expedition in the Søndre Strømfjord region to study ice sheet access and glaciological features.[25]

The Longest Trek: Project Jello

Amongst this plethora of US Cold War scientific projects in Greenland, one undertaking stands out: Project Jello, a 1955 ice sheet expedition marking "the longest trek that the United States ever did in Greenland."[26] Jello was the field designation of a six-member scientific expedition team led by US glaciologist Carl S. Benson. With Eismitte as its focal point, Project Jello was the culmination of 19 months of fieldwork spread over a four-year expedition. Cutting a north-south swath from Thule Air Base to Eismitte and beyond, the expedition provided glaciological knowledge, and hence military knowledge, of the ice sheet.

Following on the heels of the 1951 Defense of Greenland Agreement, the early-to-mid 1950s marked a period of intense US glaciological work in Greenland. From practical efforts to detect and bridge crevasses to a five-year study of pressures and deformations inside the ice sheet to the Jello expedition, US personnel launched a veritable assault on the island—one designed to strip away the secrets of the ice and render Greenland pliable to US military needs. Project Jello aimed to gain the knowledge of the physical properties of the ice sheet needed for the Air Force and Army Corps of Engineers to build, travel, and operate freely in Greenland. The expedition team's day-to-day energies and efforts focused on what can be considered basic scientific studies: snow stratigraphy, ice core research, and meteorological observations. The military dimension, however, was never far behind—and, indeed, is difficult to neatly excise from the scientists' declared purpose, "to extend our knowledge of the overall physical environment of the ice sheet."[27] The Air Force and the Corps of Engineers supported the expedition both because it promised to improve the military's capacity to operate in Greenland and because it would begin to fill the knowledge void in central Greenland—an area in which few US expeditions had previously ventured.

Project Jello formed the last part of a four-year expedition across the ice sheet—an expedition which covered 1,800 kilometers along a backwards-L shaped path from Thule Air Base in the northwestern corner of Greenland to the old French Camp VI, midway down the western coast of the island. Running through the summers of 1952–1955, the expedition as a whole conducted glaciological research at more than 400 points along the route. During the last year of the expedition, the six-member Jello team spent 12 days at Eismitte between July 21 and August 1, 1955. Traveling with four weasels and six sleds, the traverse represented the first US effort to cover long distances on the central portion of the ice sheet. The team was young, ranging in age from 21 to 31 years: Benson, the expedition leader, was still a graduate student at the California Institute of Technology; George Wallerstein, expedition navigator, was a doctoral student in astrophysics at the same school; and Alan C. Skinrood, expedition mechanic, was a master's student in mechanical engineering at Northwestern with a keen interest in stock car racing. They were joined by expedition radioman James B. Holston and assistant leader and glaciologist Richard H. Ragle. Finally, Robert W. Christie, expedition physician and tank commander veteran of the World War II Battle of the Bulge, was a first-year pathology resident at Dartmouth Medical School, where his mentor, Dr. Ralph Miller, had encouraged him to volunteer for the expedition "to the unexplored portions of the Greenland ice cap."[28]

Born in Minnesota on June 23, 1927, to immigrant parents from Sweden, expedition leader Carl S. Benson was the eldest of three sons. In 1944, as he neared the draft registration age of 18, Benson volunteered to join the Navy, but was allowed to finish high school before being sent to boot camp the following year. Discharged from the Navy in 1946, Benson took advantage of the GI Bill and began to study geology at the University of Minnesota, where he also minored in physics and mathematics. His summers were spent making extra money by working in logging camps in Washington, laying track for the Oregon and Washington railroads, and packing in a pear cannery at night. Upon completion of his degree in 1950, Benson joined the Alaskan arm of the US Geological Survey. Driven by the need to survey the vast swaths of land won in the 1848 Mexican-American War (present-day California, Nevada, New Mexico, and Utah, as well as portions of Arizona, Colorado, Kansas, Oklahoma, Texas, and Wyoming), the US Geological Survey was founded in 1879 to "classif[y] the public lands, and examin[e] the geological structure, mineral resources, and products of the national domain."[29] By the time Benson joined the Geological Survey 70 years later, the organization had its fingers in pies from cartography to mining to petroleum to water resources—anything, in short, related to the land and potentially of national economic, scientific, or political interest. Conducting mapping operations in the far north of Alaska during the spring and summer of 1950, Benson got his first field experience and his first taste of northern latitudes. His encounters with the Brooks Mountain Range, with its majestic 3,000-meter peaks and sweeping snow-studded valleys representing the northern extent of the tree line, fueled a childhood interest in the north.

Returning to the University of Minnesota for a Master's degree, Benson took up an assistantship at SIPRE, where he would work until the end of 1956. Henri Bader, who had left his position at Rutgers University to become SIPRE's chief scientist, took Benson under his wing and introduced him to Greenland. Benson first set foot on the island in 1952 when he accompanied Bader to Thule Air Base. "I was struck by the beauty of it," he recalled upon stepping off the plane: "the striking contrast of that blue water, the white icebergs and the blue sky—the blues and whites are really so outstanding."[30] "The base at Thule was truly a remarkable thing," he continued: "It was, at that time, being built at a rate that I had never imagined. They were putting up, basically, a building every day. And the base had no protection. There was no radar protection around. No anti-aircraft." His reaction echoed that of French Expéditions Polaires Françaises logistician Robert Pommier, who upon flying into the base in 1952 remarked that "the construction of Thule must have taken means comparable to those used for the 1944 Normandy invasion."[31] Concerned about Thule's vulnerability, the Air Force sent Bader and Benson out onto the ice sheet to study its basic physical characteristics—information that was needed before a radar station could be erected. "Flying over Thule Air Base, or looking at it from the surrounding hills," Benson would write later, "one often thought of it as the greatest monument to fear ever built."[32]

On his first trip to the ice sheet, while conducting measurements of temperature, melting and snow accumulation, Benson "immediately and unreservedly lost his heart to the white expanse."[33] It was the beginning of what Benson calls his "eight-year affair with the Greenland ice sheet."[34] "The polar regions have held a strong pull on a lot of people and I just happened to be one of them," he said years later—a pull that kept him far from those he loved at home.[35] In 1952, Benson had met Ruth Gronlid—"the girl of my dreams"—and the glaciologist and nurse were engaged the following year.[36] For the two years of their engagement, until the wedding in October 1955, Benson was virtually always away, and frequently in Greenland. "I am everlastingly grateful for her patience, courage, and understanding," he wrote in the preface to his 1960 PhD thesis: "In particular, her reams of letters during 24 long months of separation were indispensable."[37]

Over those two years—from 1953 to 1955—Benson split his time between several cold environments theaters, testing compacted snow runways in the Sierra Nevada mountain range, at Canada's James Bay and Labrador, and on Greenland's ice sheet. But Benson's interests lay in matters much more fundamental than runway construction: he wanted to conduct basic stratigraphic research on snow, ice and permafrost in order to understand their fundamental properties. The US Army Corps of Engineers and Air Force believed such research was essential for military operations on the ice sheet and supported Benson's plans—plans which culminated in Project Jello. Through the early-to-mid 1950s, Benson lived four years of perpetual winter working in the continental United States and in Canada during the winter months, "and then as soon as the buds would appear anywhere on a tree, it was time to go back to Greenland."[38] Benson's 19 months of fieldwork in Greenland, including Project Jello, formed the basis for his PhD dissertation, which he completed at the California Institute of Technology in 1960.

Crossing Greenland

In the context of a cross-Greenland scientific expedition, Eismitte stood out as a natural location for an elongated stopover. The data collected at Eismitte by the Wegener and Expéditions Polaires Françaises expeditions constituted the best scientific record of any location on the ice sheet, and Benson saw continued studies of the area as key to building a comprehensive view of changes in Greenland's physical environment over time. Just as the French had been eager to build on the Wegener expedition's work, so too Benson saw Ernst Sorge's early glaciological studies as providing the foundation for continued American research. For their part, the French were thrilled that Eismitte would again be the focus of scientific work: "We are overjoyed at the idea that our Station Centrale will receive visitors this year, and that our markers will be saved from being swallowed by the snow," wrote Paul-Emile Victor to Henri Bader in 1955.[39]

Unlike the German and French expeditions, the Jello team did not overwinter at Eismitte. Rather, Jello was a traveling expedition, a train of giant snow-capable weasels pulling cargo sleds and sleeping quarters, which spent a fortnight at Eismitte as part of a longer journey. The challenges of maintaining and supplying an ice sheet party over long distances and long time periods were met by the US military's heavy logistics and air presence in Greenland. The vast majority of the nearly 14 tons of equipment (including fuel, food, trail markers, and technical equipment) needed for the Jello traverse were supplied by regular air drops, allowing Benson and his men to travel as unencumbered as possible.[40]

The Project Jello team arrived at Thule Air Base on April 13, 1955, and spent a full month preparing for their ice sheet traverse and their stay at Eismitte. Much of this time was taken up by building their own wanigans and fuel hauling sleds "essentially from the runners up."[41] While the sled design came from the SIPRE headquarters in Wilmette, Illinois, actual construction took place at Thule. Of the six sleds built by the men, two were light-weight glasswool and canvas covered wagons on runners, and four were fuel- and food-hauling sleds designed to be pulled in tandem behind weasels. While at Thule, the men also planned their expedition resupply meticulously down to the last barrel of fuel, and carefully packed rations and other materials destined for air drops.

After a month of full-time preparation, the Jello weasel train set out over the vast white expanse of the ice sheet, a barren wasteland with no roads, landmarks, or other navigational aids. The series of dark dots crawling slowly across the ice was often obliterated by blowing snow and blizzards which could bury the vehicles entirely in a matter of hours. The living quarters were small, "seldom warm by ordinary standards, and often damp."[42] In the evenings, a gas stove provided a modicum of heat and lanterns allowed the men to read, repair equipment, plot data, and socialize. To avoid condensation in the vehicles, cooking took place outside in a tent insulated with cardboard and fiberglass. Fearful of fires and carbon monoxide poisoning, the men turned off the heat overnight, choosing instead to burrow inside sleeping bags arranged in the wanigans—sleeping bags that they often shared with frozen food, brought inside to thaw for the following day. Apart from the discomfort of cooking in frequently inclement weather, the rations provided to the Jello expedition were both varied and ultramodern—a deliberate move to increase team morale and decrease the weight of supplies. Armed with a two-burner camp stove, a small stove-top oven, a pressure cooker, and assorted pots and pans, the men took it in rotation to cook meals including, at their best, "fine turkey dinners with cranberries, green beans, and mashed potatoes with gravy or butter."[43]

From the very conception of Project Jello, Benson insisted on the need to keep ration weight as low as possible. "One of the things we wanted was lighter weight food," he said: "The Army C-ration runs 6.4 pounds per man day. This is a number I'll never forget. And the 5 in 1 rations run just about 6 pounds per man day. And I wanted to get our weight down to 2 pounds per man day."[44] Pointing to the logistical troubles of SIPRE's 1954 Party Crystal expedition, which conducted snow studies in northern Greenland but struggled under the weight of giant 46-ounce juice cans and other unwieldy rations, Benson argued that "the large food requirement of 7.4 pounds per man per day had to be reduced."[45]

Modern rations research and development in the United States dated from 1936 when, under the impending threat of war, the Quartermaster Subsistence Research and Development Laboratory was tasked with developing improved rations.[46] Through World War II, the Quartermaster Laboratory sustained US forces with newly designed C-rations and 5-in-1 rations. A daily food and accessories ration for soldiers cut off from regular supplies, the C-ration consisted of a meat and vegetable meal as well as cigarettes, water purification tablets, matches, toilet paper, chewing gum, and a can opener. The 5-in-1 ration was more specialized, designed for combat troops in remote desert areas. Benson, however, wasn't impressed by either option. "Much of their cargo-weight is not eatable (such as tin cans, cardboard boxes, and wire) or unnecessary, such as the large quantity of fluid in canned foods," he wrote, concluding that existing rations contained "too many luxuries": "It's a banquet in a can. It's too expensive. We can't afford to carry it."[47]

Working with the Quartermaster Food and Container Institute (as it was renamed in 1946) in Chicago, the Jello team procured new dehydrated foods and banished all unnecessary water and packaging from the rations. The result was a menu of frozen, dehydrated and crystallized foods: lightweight, compact, and easy to restore with plentiful access to water from melted snow. The adoption of new food technologies such as grapefruit and orange juice crystals, Benson wrote in his final report on Project Jello, "increased our range of travel and permitted greater independence from time consuming air drops."[48] The difference made by the new foods was striking: Project Jello carried just over half the ration weight of other US Army projects in Greenland, resulting in a savings of nearly two tons. "For a small, mobile expedition weight considerations are extremely important," emphasized Benson and assistant leader Richard H. Ragle: "Fuel weights for a given distance cannot be easily altered, but food cargo weight can be altered by careful choice of dehydrated and frozen foods."[49]

As to their taste, variety, and ease of preparation—also important elements in the planning process—the rations received mixed reviews from the Jello team. Variety was a success: in a far cry from early twentieth-century polar expeditions, the men traversing Greenland's ice sheet in the summer of 1955 could choose between oven roasts, veal loaf, and ground beef; sweet potatoes, green beans, and cabbage; and lima bean soup, eggs, and canned cheese. By far the most popular item, Benson wrote immediately upon return, were the T-bone and porterhouse steaks. "The steaks were used at least once each week—for breakfast as well as for dinner—as long as they lasted," he reported: "They were completely gone when the trip was two-thirds over."[50] Other cuts of meat, however, impeded on the day-to-day travel and operations of the group. "The roasts were cooked in an oven after a thaw period from 6 to 8 hours," wrote Ragle: "Usually the thaw period was not sufficient [and the] cooking process took from 3 to 4 hours in a medium oven. Texture and flavor were excellent [but] the time taken to prepare a roast was more than was generally available." The dehydrated vegetables, too, came in for criticism: the sweet potatoes, complained physician Robert W. Christie and mechanic Alan C. Skinrood, were a "poor item, not at all popular" with "virtually nil" flavor.[51] The green beans, on the other hand, stood out as "the only green we saw in one hundred days" and, as such, were "a good psychological point where most items looked pretty drab." In the end, the rations delivered satisfactorily on both nourishment and palatability, to the extent that the expedition's radioman gained five pounds that summer.

Ice Sheet Logistics

The logistics of a long traverse on the ice sheet proved a significant challenge to Jello's planners. With a 8,000 kilometer perimeter "rimmed by a steep-sloped belt of rough jagged ice, 10 to 50 miles wide, and so broken by dangerous crevasses and scored by deep ice channels that it is practically impassable," as a US military document described it, Greenland's ice sheet can only be accessed overland by load-bearing vehicles at a handful of locations.[52] Supplying the Jello expedition by land was thus a near impossibility given the sheer quantities and weights of the materials necessary to support a traveling party of six men with a robust scientific agenda for over three months. The supplies in question included more than 9,000 kilograms of fuel, 1,000 kilograms of bamboo and plywood for snow accumulation markers, and food rations to sustain the men in cold temperatures and at high altitudes.[53] Despite the challenges posed by overland supply routes, and despite successful French use of air drops during the Expéditions Polaires Françaises expedition to Eismitte, US planners saw air support as unsatisfactory and overly expensive as late as 1953. "Access to the Ice Cap by airlift has heretofore not been considered practicable," asserted the Arctic, Desert and Tropic Information Center in that year: inability to deliver delicate equipment and personnel by air drop, limitations on the cargo capacity of available aircraft, and the dangers of ice sheet flying were all listed as reasons not to pursue air support.[54]

The "cumbersome and complex" logistical problems of Benson's planned 1955 trek, however, necessitated a fundamental rethink of these attitudes.[55] Despite their initial misgivings, US planners eventually saw air support as the only hope for a successful traverse. A medley of US military groups, including the Northeast Air Command, the

First Engineer Arctic Task Force, the Army's Transportation Arctic Group, and the 6614 Air Transport Group, worked together to provide air support to Project Jello out of the Thule and Sondrestrom Air Bases. The supplies prepared by Benson and his men, including large quantities of fuel, were free dropped by low-flying aircraft along the length of the traverse. Prior to each airdrop, the Jello team staked out a drop zone parallel to the wind using everything at their disposition to guide the pilot: colored flags, smoke bombs, and even the weasels and tents themselves were enlisted to indicate the boundaries of the drop zone, the height above the ground, and the wind conditions at play.[56] Between setting out the drop zone, guiding the pilot, digging the supplies out of the snow, and loading the weasels, each airdrop consumed an entire day's work.

From the ground, the airdrops were breathtaking and "beautiful to watch," but they were fraught with danger for the pilots.[57] In the best of circumstances, ice-sheet flying was a risky undertaking: with no landmarks and unpredictable and often ferocious weather conditions, the vast ice sheet did not leave room for error. For the Project Jello airdrops, the pilots had to navigate their aircraft to an altitude of five to ten meters in order to prevent supply barrels from breaking, bursting, or impaling themselves irretrievably in the snow. Under these conditions, the pilots had little ability to judge their course, and, indeed, the men on the ground were better able to judge the aircraft's altitude than the pilot himself. First, simply locating a small group of vehicles and men on the ice sheet presented difficulties regardless of the weather. The clear visual establishment of a drop zone marked out by as many objects as possible was essential since, as Benson wrote in a 1955 report on the resupply of ice sheet expeditions by air, "there are no other features which can be used for depth perception and estimating distances on the surface."[58] Constant radio communication allowed the men on the ground to guide the aircraft and abort the drop if necessary: "being in radio contact with them on the ground, we could see and judge their altitude better than they could often judge it while flying. So, they would depend on us to say, as we did have to tell them several times, just abort the drop and climb as fast as you can and get out of here. And then they'd make another run and do it at the proper altitude. It couldn't be too high and it couldn't be too low."[59] Despite these dangers, and despite the initial reluctance of US planners to authorize air support, "the entire air drop resupply operation worked extremely well and 100% recovery was obtained from all fuel and food drops."[60] And the men on the ice sheet appreciated the drops for more than just fuel, food, and scientific equipment: they also brought a connection to home in the form of regular care packages. "It was a big deal to get an air drop," recalls Benson, "[because] that's when you get your mail. [My fiancée] was very good at writing letters and sending cookies and doing things like this."[61] At each airdrop, he continued, "I'd get a big wad of letters and I would line them up according to postmark and then read usually one a day or a couple a day. It depended on how impatient [I'd] get. But, I liked to piece them out because I knew we weren't getting any other mail for a long time."

Project Jello's airdrops were based on the Expéditions Polaires Françaises system. Benson repeatedly emphasized his debt to the French, highly recommending Paul-Emile Victor's expertise on airdrops to those considering their own Arctic expeditions.[62] This connection was representative of one side of an ongoing but unsettled relationship between US polar operations and Victor's Expéditions Polaires Françaises. The Army's SIPRE establishment, including Benson and Bader, thought highly of Victor, repeatedly providing generous logistical assistance to French efforts in Greenland through the 1950s and 1960s. And in 1952, only a year after the end of EPF's second overwinter at Eismitte, SIPRE recognized the French organization's expertise in ice thickness measurements and contracted EPF to extend seismic measurements to northern Greenland.[63] In the summer of that year, a joint team of six Frenchmen and five Americans spent more than two months traveling 2,000 kilometers across the ice sheet from Thule Air Base to Danmarksfjord, conducting seismic measurements at 24-kilometer intervals. On the other side of the coin, however, the State Department took a neutral and pragmatic attitude towards EPF, agreeing to Victor's frequent requests when it suited them and otherwise turning him down with little apology.[64] And the US Air Force was less than impressed with the dean of French polar science, all but cutting him out after an unsuccessful joint mission in 1953. This incident dealt a harsh blow to Victor's pride, especially since he still felt a strong connection to the Air Force, for whom he had worked during World War II, but underlines the essential asymmetry of the situation: to the United States, Victor was a consultant who could be called upon as needed, but to whom the United States owed little, while to Victor, the United States represented a deep emotional investment.[65]

Project Jello Arrives at Eismitte

Out of the three months of the Jello traverse, the fortnight at Eismitte was the clear highlight, both personally and scientifically, for the team. "Our arrival at the Central Station was one of the most thrilling events in my life," Benson wrote to Victor in October 1955, and the French, too, were overjoyed at having their station once again at the center of scientific inquiry.[66] "I have just learned that the SIPRE research group has refound our Station Centrale in Greenland," wrote French glaciologist Albert Bauer to Henri Bader only three weeks after the Jello expedition's arrival: "I congratulate you, as well as Mr. Benson, for this feat."[67]

Finding the old French station, however, was far from simple: in the four years since the French had closed the camp, the buildings had been covered by more than five meters of snow. Anticipating difficulties, Bader wrote to Victor in the spring of 1955 asking for advice on finding the station, locating the French snow markers, and accessing the buried buildings themselves.[68] Because of the late timing of the letter, sent just two weeks before Benson and his team were set to leave Thule, Victor's reply had to be parachuted to Benson on the ice sheet—fortunately for all concerned, Victor responded in English. The response included detailed locations for the nearly six-meter long bamboo sticks marking Station Centrale's trapdoor, as well as the most recent French photographs of the area.[69] Victor also provided directions to find a food cache, buried near Station Centrale's main entrance—a cache, which the French had left in the hope that other expeditions might follow in their footsteps. With this help, Benson and his team found the French camp by surface exploration, even though previous aerial reconnaissance flights had failed to spot the site. By excavating several meters of snow from the entrance, Benson was able to enter the partially crushed station, marking the first human presence in the French camp for four years. The US team also unearthed the French food cache and was greatly impressed by the rich flavor and creamy consistency of the French powdered milk—a far cry from the thin and lumpy American variety.[70] In order to ensure that the station remain findable in future years, Benson's team erected a nine meter tall steel tower close to the principal entrance, approximately two-thirds of which extended above the 1955 surface of the snow.[71]

Upon arrival at Eismitte, Benson's team quickly got to work on drilling a 16-meter deep ice core—an endeavor that both followed directly in the footsteps of the German and French expeditions and was representative of the project's scientific agenda. Project Jello's Eismitte ice core was a cornerstone in a work plan that saw Benson and his team dig and drill into the ice at nearly 150 sites over four summers. They were inspired by Ernst Sorge's demonstration that annual layers could be identified through density differences—work that Benson referred to admiringly as a "classical study" in glaciology.[72]

Dug by hand, Benson's pits ranged from three to six meters deep—an undertaking, which, during the 1955 Jello expedition alone, involved moving over 600 tons of snow.[73] At most stations, the pits were extended downwards to 11 meters with an ice core drill. Developed by the US Army Arctic Construction and Frost Effects Laboratory and modified by SIPRE, the drill was a reliable if crude hand auger. The drilling work was slow and difficult, averaging an hour per meter and often hampered by the weather: on windy days, when snow could fill the pit in a matter of hours, work took place in the dark under a makeshift roof. The ice core samples were then extracted, described, cut, measured, and weighed—a time-consuming process that typically took two days. In each pit, Benson's team also measured temperature, density, hardness, and grain size in the visible strata, through which they could determine annual accumulation. The sheer scale and physical demands of the work led Chester C. Langway, US geologist and pioneering ice core researcher, to praise Benson's ice core work as a "Herculean research effort."[74]

At 16 meters, the Eismitte ice core was one of the deepest of the entire expedition. While it was not as deep as the core recovered by the French five years earlier, it was of better quality, allowing for accurate analysis. Project Jello's best ice cores, including that from Eismitte, were carefully packed and sent to geochemist Samuel Epstein at California Institute of Technology. Born in Poland and raised in Canada, Epstein worked at the Canadian Atomic Energy Project before moving to the United States to work under Nobel laureate Harold Urey. As a young researcher at the University of Chicago, he became interested in the analysis of oxygen isotopes in natural materials. Along with his Chicago colleagues, Epstein experimented with determining ancient ocean temperatures from measurements of the oxygen-18 to oxygen-16 isotope ratios in geological samples. After moving to California in 1952, Epstein turned his attention to ice, first from Alaska's glaciers and then from the Jello expedition's Greenland ice cores. By subjecting ice samples to oxygen isotope analysis, he was able to uncover detailed annual gradations.[75] Epstein's analysis provided a level of precision above Sorge's density measurements of a quarter-century before, giving, as Benson wrote admiringly, "a beautiful representation of the annual units."[76] This proof of concept stimulated the US to invest heavily in ice core research both within military contexts and as part of the 1957–1958 International Geophysical Year (and, later, through the National Science Foundation). It was, Langway continued, the beginning of a now standard and critically valuable method of understanding past climates through ice cores.[77]

At the same time as Project Jello's Eismitte ice core was under the microscope in California, a young Danish researcher, Willi Dansgaard, was also working with oxygen-18 to oxygen-16 ratios. Trained as a physicist, Dansgaard worked briefly for the Danish weather service in the immediate postwar period but soon left for the University of Copenhagen's Biophysical Laboratory, which offered an environment more conducive to individual research. It was at the Biophysical Laboratory that Dansgaard began thinking about oxygen ratios and their analytic potential, working with rain and river water samples provided by the Danish East Asiatic Company's worldwide network and cloud water vapor samples he and his wife collected themselves on a Royal Danish Air Force flight. In 1954, he developed the idea that the oxygen isotope ratio in old water could give information about the climate at the time that water was formed—the key link between Greenland's ancient ice, core drilling, and past climates. Stimulated by the quantitative nature of Epstein's work, Dansgaard was eager to study Greenland's ice himself. Dansgaard's subsequent work, taken up in the next chapter, would unlock paleoclimatological secrets from Greenland's ice and carve a new place for Denmark in glaciological research.[78]

A Fine Balance: Denmark, Greenland, and the United States

Project Jello was conducted nearly autonomously from any Danish involvement or participation—a situation not unusual for US scientific work in Greenland at the time. Still, while no Danish scientists or minders accompanied Benson's team, and while no Danes were involved in the planning of the scientific work, the island's political situation was still intimately connected to the expedition.

With the United States' all but limitless military access to Greenland during World War II weighing heavily on their minds, the Danish postwar government—newly reassembled on May 5, 1945 following five years of German occupation of metropolitan Denmark—was deeply concerned about regaining and maintaining control over the island in the opening years of the Cold War.[79] As a small state caught between two emerging superpowers, it quickly became clear that Denmark's traditional stance of neutrality and nonalignment was no longer necessarily in the country's best interest. In Danish political circles, no topic was more pressing than bolstering security and demonstrating sovereignty in the new geopolitical framework. As the 1940s turned into the 1950s, Greenland's natural isolation was pierced not only by US military interests, but also by groups as varied as French fishermen, British scientists, and eager foreign prospectors, as well as regular passenger flights from Western Europe to North America that stopped for refueling in Søndre Strømfjord.[80] "The Greenland which lay remote from the civilized world for thousands of years has become, in less than a mere decade, a public thoroughfare," proclaimed the Danish Foreign Office Journal, noting that by 1956 Søndre Strømfjord boasted a hotel that served its guests whisky poured over inland ice.[81] These multitudinous incursions redoubled Danish determination to exert sovereignty over Greenland.

As tensions mounted across the top of the world and US calls to secure Greenland grew louder, the Danish government became increasingly reluctant to negotiate directly with the dominant country.[82] "While we owe much to America," said Danish Foreign Minister Gustav Rasmussen in a telling moment, "I do not feel that we owe them the whole island of Greenland."[83] Still, the government also recognized its own inability to provide full military security for the strategically important island. In this context, the North Atlantic Treaty offered Denmark the opportunity to conduct negotiations over Greenland in a broader setting—an opportunity that made the embryonic alliance appealing to the small country despite its traditional nonalignment. Rasmussen used precisely this argument to win over skeptical members of parliament: "the alliance would provide a multilateral setting for specific defense questions such as the future use of Greenland for Atlantic defense," he asserted in a speech to the Danish *Folketinget* in March 1949.[84] Rasmussen's arguments were persuasive, and the parliament voted overwhelmingly to become one of 12 founding members of the new military alliance.[85] On the US side, negotiators were fully aware of the political situation in Denmark. "Danish public opinion is decidedly adverse to any diminution in Denmark's sovereignty over the last remnant of her overseas empire, which, it is feared, would be the inevitable consequence of a prolonged American occupation," asserted a confidential US military report: and, further, "Denmark fears the policy of granting concessions to one of two great powers in an area vitally important to security of both will invite retaliatory measures from the other."[86]

Following NATO-backed negotiations, Denmark and the United States signed the Defense of Greenland Agreement on April 27, 1951. The agreement emphasized Danish sovereignty over the island while also ensuring its defense through "arrangements under which armed forces of the parties to the North Atlantic Treaty Organization may make use of facilities in Greenland in defense of Greenland and the rest of the North Atlantic Treaty area."[87] The United States was granted exclusive jurisdiction over three defense areas at Thule, Narssarsuaq, and Søndre Strømfjord. Outside the defense areas, however, all US activities required formal Danish permission. On an annual cycle, the US embassy in Copenhagen submitted upcoming projects to the Danish Ministry of Foreign Affairs, which consulted as needed with the Ministry of Defense, the Greenland Ministry, and the Parliamentary Foreign Affairs Committee before granting or refusing permission.[88] This procedure, cumbersome and time-consuming to US eyes, was vital to Danish interests: retaining effective sovereignty over the island, placating public opinion, informing themselves of US activities, and protecting Greenland's indigenous peoples.[89] Still, lack of clarity and confusion—both intentional and unintentional—abounded.

The Danish authorities, as historian Nikolaj Petersen has emphasized, "did not notice any clear example of unauthorized US research activities, but they often found it difficult to determine whether the actual projects were covered by the given permissions—the more so because permissions had to be given on the basis of frequently unclear applications."[90] Even after a Danish scientific advisor was posted to Thule in 1954 in order to keep a closer eye on US activity, the dispersion of this activity across the island made it increasingly difficult to track. Aksel Nørvang, the second scientific advisor at Thule and a geologist by training, worried that the scope and scale of US research in Greenland made it hard to grasp this work in its entirety, and suggested that Denmark was not being fully informed about US research activities.[91] "In many cases, it is difficult to determine whether the [US] scientific work had any bearing on military affairs," agreed the Greenland Ministry's Permanent Secretary Eske Brun in a comment that reflected the Danish unease over US military activity in Greenland.[92] The Danish Foreign Ministry, too, believed that "on the whole, it can probably be said that US studies in Greenland are inadequately followed by the Danish authorities"—an unsatisfactory situation caused in part by personnel shortages.[93]

On the US side, care was taken to follow the letter, but not always the spirit, of the permissions procedure. Acutely aware of sensitivities over US presence on the island, diplomats at the US embassy in Copenhagen repeatedly urged the State Department to treat Greenland carefully. "As the Embassy has pointed out on different occasions, there is a certain sensitivity among the Danes with respect to the sovereignty of this almost last-remaining Danish territory beyond metropolitan Denmark, and this occasionally shows itself, as when the United States operations on Greenland from time to time go beyond expressly-granted rights, even though these deviations are only slight and accidental," the US ambassador in Copenhagen, Robert Coe, cabled to the State Department in 1957.[94] "One of the principal objectives of our diplomacy in Denmark," Coe continued, "[is the] maintenance of an attitude favorable for continued use of the bases and continued cooperation with respect to their use." The embassy played a delicate balancing game, negotiating regularly with the Danish government on topics including flight permits, radio frequencies, the use of Danish contractors, and research and operations outside the defense areas. But the information provided by the Americans was also carefully constructed so as to allow the Danes only certain insights into US work in Greenland: oftentimes, little detail was given about military operations and scientific studies, and lead times were very short. For more sensitive projects, the true nature of the work was shielded from Danish eyes. In Operation PCA 68 (a rocket project to study effect of a nuclear disturbed ionosphere on radio communications), as historian Henrik Knudsen has shown, "the military agenda and the military background knowledge were inaccessible and remained unknown to almost all agents outside the US"—and during negotiations with the Danish government, the US side "blur[red] the link to military sponsorship [to] render it more acceptable."[95] More dramatically, the Danish government had no notion of Project Iceworm, an aborted 1960s effort to deploy 600 nuclear missiles on railroads under Greenland's ice sheet, despite preliminary studies conducted on the island.[96]

Conducted as it was well outside the defense areas, Project Jello required preapproval from the Danish authorities. For Benson, this proved a source of great frustration. In the summers of both 1953 and 1954, Benson's planned expeditions on the ice sheet ran into the political wall: in 1953, US authorities forgot to ask for clearance and the expedition had to be curtailed, and in 1954 an apparent mix-up with permissions meant that the Danes did not grant clearance to Benson's team, again forcing changes to the workplan.[97] At the start of the Jello traverse in 1955, Benson put in a personal appeal for political clearance to be handled expediently. When he arrived at Thule in April of that year, however, Danish clearance for the expedition had still not come through—apparently because the US side had not included Project Jello in its annual list of projects.[98] Rather than letting the expedition be affected for a third year in a row, Benson and his navigator, George Wallerstein, took the matter into their own hands: they followed their pre-established expedition route, but always radioed in positions that were in the cleared area—all the while planning to blame any disputes on navigation errors.[99] In the end, the

Danes did not question the expedition and hence no creative navigation narrative was necessary. The story, however, shows how little control Denmark had over the vast island—a situation that would, in the 1960s, contribute to a Danish drive to take ownership of glaciology on Greenland's ice sheet.

A Scientific Cold War in Greenland

The US Cold War presence in Greenland represented the first time that regular scientific work was conducted on the ice sheet: from even before the ratification of the 1951 Defense of Greenland Agreement onwards, American military and scientific personnel were present and working in Greenland without interruption. From the permanent installations at Thule and Sondrestrom Air Bases to temporary facilities at Site 2 and Camp Century to roving ice sheet expeditions, US presence on the island was so great that it has been described as a hegemonic colonization of Greenland.[100] Indeed, the sheer volume of US personnel and equipment in Greenland far exceeded any previous work on the island: Thule Air Base was the largest US Air Force base outside the United States, with a postconstruction population peaking at 10,000 men in 1962.[101] The existence of this facility in northwestern Greenland, with its 3,000 meter runway, hangers that could fit giant Lockheed C-141 Starlifter aircraft with ample room to spare, and full amenities for military and civilian staff, enabled US personnel to support year-round operations all over the island. Thule was, in fact, a veritable city, boasting a radio station and cinema, a permanent chaplain, recreation facilities, and even a Christmas tour stop from the Bob Hope Show. The emergence of a mode of exploration that was regular, long-lived, and fully supported within Greenland itself was a product of Cold War realities: this was exploration guided by explicit military aims, and supported by military infrastructures. It is precisely this level of operation that prompted Danish glaciologist Børge Fristrup to go as far as saying that, with the US work on the ice sheet, "the age of real exploration is passed in Greenland... It has been replaced by a no less exciting program of technical and scientific research."[102]

Military and civilian scientists, including Carl S. Benson, were supported not only by military funding but also by an entire infrastructure built to master the challenges of the ice sheet. At the SIPRE headquarters in Wilmette, Illinois, the CRREL headquarters in Hanover, New Hampshire, and the Natick Laboratories in Massachusetts, thousands of US personnel worked stateside to assist missions in Greenland and other northern theaters. New facilities were built to design and test vehicles for transport over snow and ice, clothing and footwear for soldiers and other personnel in polar regions, and even a portable nuclear reactor for use on the ice sheet.[103] The resources devoted to the US mission in Greenland were a product of perceived Cold War strategic necessity. In the US tradition of Big Science, large-scale investment in science and technology, coupled with large budgets, was seen as the key to conquering the ice sheet. For civilian scientists on the ground in Greenland, the entire system seemed both opaque and hugely generous: "It was like having an infinite budget," Benson said in 2001: "The program was overwhelmingly big. I just can't get over that. When we look at the way we budget things today, and think back to how things were going then, there was never a nickel exchanged. There was never talk of money. If we needed Weasels, if we needed an aircraft, we just put in the request and we got these things."[104] Benson's Project Jello expedition to Eismitte was backed by a behemoth: the US defense establishment, which saw military control and scientific understanding of Greenland's ice sheet as critical to North American continental security. In this context, Eismitte was transformed into a laboratory of the Cold War.

Chapter 4

It Has Completely Changed

A European Cooperation: Expédition Glaciologique Internationale au Groënland

In the early postwar years, disciplinary growth in glaciology quickened. Across Europe, new glaciological societies and research bodies were formed and existing ones strengthened. Among others, the Swiss Commission for Snow and Avalanche Research, dating from the nineteenth century, built a new, modern snow and ice research laboratory immediately after the war, the French Hydrotechnical Society founded a glaciological section in 1948 to bolster the study of French glaciers, and the Icelandic Glaciological Society, established in 1950, built a research station on the Vatnajökull ice dome in the same year. And in Norway, when the *Norsk Polarinstitutt* (Norwegian Polar Institute) brought world-renowned oceanographer Harald Ulrik Sverdrup home from the Scripps Institute for Oceanography in La Jolla, California, to head the Norwegian institute, one of Sverdrup's first moves was to add a glaciologist to his staff in order to build a program of glaciological research in Norway and on Svalbard.[1]

The trend towards stronger professional and scientific frameworks for glaciology was not only a European phenomenon: in North America, the rising importance of the Arctic to sovereignty and security considerations led to the creation of the Arctic Institute of North America (AINA), a private binational body, in 1945.[2] Headquartered at Montreal's McGill University and supported financially by organizations as varied as the Carnegie Corporation, the Office of Naval Research, and the Northwest Territories Council, AINA was an important postwar sponsor of Arctic-oriented scientific research across Alaska, the Yukon, and the Northwest Territories. As well as instigating studies of glacial systems, snow dynamics, and ice cores,

AINA won research contracts from US military bodies and served as a polar think tank. The postwar years also saw increased cooperation among US bodies with glaciological interests, including the American Geographical Society, the American Alpine Club, and the US Geological Survey. And in Canada, the National Research Council established a specialized research section to develop Canadian solutions to snow and ice problems ranging from the passage of military ships through Canada's Arctic archipelago to snow clearing on Rocky Mountain railways to polar travel techniques for the Royal Canadian Mounted Police.[3]

International scientific bodies, too, grew to link the glaciological world. In 1947, the International Glaciological Society (itself dating from 1936) founded a new periodical, the *Journal of Glaciology*, to help cope with the rise in interest in glaciological research stemming from the war. With regular contributions from across the United Kingdom, continental Europe, and North America, the journal proved an important mechanism for communication and bridge building in a growing discipline. Further, the journal's wide scope meant that it quickly became a discussion ground for professional and amateur glaciologists; for academic, private, and military glaciologists; and for interested members of related disciplines from meteorology to rheology to geography. All agreed that "long-term co-ordinated field, laboratory and theoretical research," as Henri Bader put it at a glaciological conference in New York in 1949, was vital.[4] The early postwar years also saw the International Commission of Snow and Ice emerge as an independent body from under the umbrella of the International Association of Scientific Hydrology. Led in its first years by glaciologist and soon-to-be Swedish ambassador to Norway Hans W. Ahlmann, the International Commission of Snow and Ice's symposia provided regular and dynamic meeting places for glaciologists from as far afield as Europe, North America, Japan, India, and South Africa.[5] It was this last organization that would set in motion the next major scientific expedition to Eismitte.

European Cooperation in Greenland

At its Rome meeting in 1954, the International Commission of Snow and Ice announced a new expedition to Greenland.[6] From the first instance, this was designed to be a large-scale European cooperation. With its glaciological focus, the expedition attracted the attention of five nations with long-standing interests in that field: France, Switzerland, Austria, the Federal Republic of Germany, and Denmark.[7]

By combining expertise and resources, the hope was to spread the high costs of a polar expedition over a number of countries and to make the best use of diverse research expertise. The new expedition, known by its French title *Expédition Glaciologique Internationale au Groënland* (EGIG), was scheduled to run from 1957 until 1960, coinciding with the 1957–1958 International Geophysical Year (IGY).

The primary scientific aim of the EGIG expedition was to expand on previous studies in central Greenland, and specifically at Eismitte, in order to "resolve the geophysical enigma represented by the ice sheet."[8] Both the timeframe and the geographic scale were ambitious: EGIG aimed to study the horizontal band of the ice sheet delineated by the 68th and 73rd parallels, stretching nearly 1,000 kilometers from coast to coast. Within this band, which represents Greenland's area of greatest ice activity, two locations were singled out for extensive glaciological studies: Eismitte, important because of its long history of scientific research, and Jarl-Joset Station, near where Frenchman Alain Joset and Dane Jens Jarl had perished in 1951. Building on the work of earlier expeditions, EGIG planned to drill ice cores for physical and chemical analysis, to conduct seismic surveys in order to improve the subsurface map of Greenland, to study the mass balance and surface velocity of the central ice sheet, and to establish a wintering station. In line with a growing trend of the era, EGIG's scientists hoped that their work would "deliver...information of strong paleoclimatological interest."[9] While these scientific goals were central to the planning of the expedition, the involvement of five countries with different (and, at times, competing) interests meant that politics was never far behind.

As the realities of organizing a cooperative European expedition to Greenland set in, a division of labor fell in place between the participating countries. France, represented by Paul-Emile Victor and 25 members of his Expéditions Polaires Françaises (EPF) team, took the helm.[10] France was considered the only participating country with "great experience [in] the *organization* of intensive scientific research" in the polar regions and the only country capable of furnishing the necessary weasels, aircraft, and other equipment—a legacy of EPF's expedition to Eismitte, which gave Victor "a primordial place in the scientific and polar worlds."[11] The scientific work itself was split between the five countries. Germany took charge of geophysical research on the ice sheet in what represented the country's first return to the polar regions since World War II. Switzerland took primary responsibility for glaciology on the ice sheet (including ice core research), Austria for measuring heat balance and radiation, and

France for diverse glaciological, seismological, and hydrological work. The Danes claimed the smallest slice of research, restricting themselves to glaciological and geodesic studies on Greenland's coasts. Finally, the United States, with its large presence on and interest in the ice sheet, provided logistical, technical, and moral support to the expedition.

Even as EGIG promoted itself as a robust five-country partnership, two members stood out in terms of leadership, decision making, and financial commitment. Having accepted "with pleasure" to lead the expedition, France—and specifically Paul-Emile Victor—exerted by far the most influence over EGIG's logistics and scientific agenda.[12] With his connections in French diplomatic circles, Victor was able to call in official help when needed, including getting the French Ministry of Foreign Affairs to pressure the Danes into consenting to the expedition and the French embassy in Washington to lobby for American support.[13] French influence over day-to-day decisions and politicking was also boosted by Albert Bauer's role as EGIG's secretary. With his extensive polar and glaciological experience, including as an EPF veteran, vice president of the International Commission of Snow and Ice, and president of the French glaciology committee for the IGY, Bauer was a sensible choice for the post of secretary. Taking a sabbatical from his position at the *Ecole Nationale Supérieure des Arts et Métiers* in Strasbourg, Bauer committed himself fully to EGIG. His perfect command of French and German, together with his decent English, meant that he could communicate with participants from all EGIG member countries, and his "extremely dynamic and stubborn" personality was no small asset in building consensus among the participating nations—and, on occasion, preventing the entire project from falling to pieces.[14] Switzerland also played a leading role, with the Swiss president of the International Commission of Snow and Ice, Richard Haefeli, presiding as EGIG's chief scientist.

Financially, EGIG relied heavily on the French. In continuation of the government support for polar research first expressed in EPF's inaugural campaign to Greenland, France's Minister of Economy and Finance, Pierre Pflimlin, agreed to provide an extraordinary subvention to cover more than two-fifths of EGIG's mammoth 500 million franc budget (USD $12 million in 2012 dollars).[15] The monies came from the same trio of governmental bodies that had underwritten Victor's expeditions for nearly a decade: the Ministry of Economy and Finance, the National Center for Scientific Research, and the Ministry of National Education. With Charles de Gaulle's return to the Elysée Palace in 1958, French science entered a new golden age: to the President, scientific research was a driver of economic growth, social progress, and national prestige. The National Center for Scientific Research benefited handsomely, seeing its budget double between 1959 and 1962, and generously supported Victor's polar efforts.[16] Combined with the French Armée de l'Air's contribution of airplanes, helicopters, and pilots, this support amounted to more than half of EGIG's needs.[17]

The German, Swiss, Austrian, and Danish governments also contributed financially to the expedition by equipping and supporting national scientific teams.[18] On the Swiss side, additional funds came from the Helvetic Society for the Natural Sciences, the Swiss Foundation for Alpine Research, and the Academic Alpine Club of Zurich. The Danes limited themselves to a boat for geodesic studies, the *Ole Rømer*, minimal financial resources, and a small number of participants. Of the more than 40 EGIG participants on Greenland's ice sheet in 1959, only one was a Dane—and this was a government-assigned observer.[19] No Danes contributed to scientific work on the ice sheet, a situation that caused resentment among the other participating countries and embarrassment in the Danish scientific community. Danish ice core researcher Willi Dansgaard went so far as to call his country a "deadhead... whose 'contribution' was little more than permitting the project and requiring all scientific papers to be published in the old Danish periodical Meddelelser om Grønland—at the author's own expense!"[20] Helge Larsen, director of the Arktisk Institut and curator of the Danish National Museum's ethnographic collection, agreed, complaining at an internal meeting that "other countries have done enormous amounts of work [and] [f]or Denmark to participate on an equal footing with the other countries, it should contribute more than just sending representatives to the meetings."[21] In the early 1960s, these sentiments would be swept up in a broader call for Danish authority over glaciology in Greenland.

Even the United States, not officially an EGIG partner, provided more than the host Danes, including air support worth 75 million francs (USD $1.2 million in 2012 dollars).[22] Championed by US glaciologist and scientific administrator Henri Bader, US support for EGIG was connected to the fallout from the IGY of 1957–1958. The US IGY committee is "completely dominated by the upper atmosphere people," an irritated Bader wrote to Bauer in a "very confidential" note.[23] Bader's frustration was underscored by the funding of the US IGY glaciological program, whose original budget of USD $1.4 million was cut by more than half to USD $650,000 before the geophysical year even got underway—a blow to Bader's desire to use the IGY as a launchpad for US ice core research.[24] Pushing back against what he saw as the US IGY committee's "very negative" attitude towards glaciology, Bader offered EGIG "greater SIPRE support than you might have anticipated," including air support, instruments, and weasels. "We have a vital interest in Greenland research and will use all our power, which is fortunately independent of IGY to help you get into the field and work there," he assured Bauer.

EGIG was in many ways a microcosm of the IGY, occurring simultaneously and espousing the same values: international scientific cooperation, synoptic data collection, and a vision of the earth as an integrated system. Following in the footsteps of the International Polar Years of 1882–1883 and 1932–1933, the IGY of 1957–1958 saw tens of thousands of scientists from nearly 70 countries participating in research centered on the earth and geophysical sciences.[25] Taking place at the height of the 11-year sunspot activity cycle, the IGY balanced calls for scientific internationalism with the competing national pursuits propelled by the ongoing Cold War and further ignited by the launch of Sputnik in October 1957. In the polar regions, both western nations and the USSR undertook glaciological research, ice core drilling, and seismological measurements of ice thickness.[26] The connections between EGIG and the IGY were more than simply a matter of shared timing and values: EGIG's glaciological data was incorporated into the IGY World Data Centers, new facilities designed to collect, preserve, manage, and distribute the reams of data resulting from the geophysical year, and located on both sides of the Iron Curtain.[27] And the French polar outfit, EPF, participated directly in the IGY in Terre Adélie, conducting year-round observations at Base Dumont d'Urville with research funding from the French government and a ship provided by the IGY.[28]

The first three years of EGIG, from 1956 to 1959, were devoted to preparation and preliminary exploration, including helicopter reconnaissance missions and supply caching. The principal scientific campaign took place in the spring and summer of 1959 with more than 40 scientists working on Greenland's ice sheet and coastal zones. They were supported by several dozen EPF technicians and helicopter pilots, as well as nearly 60 members of an airdrop supply team, for a total of 152 people on the ground in Greenland—by far the largest civilian scientific expedition to explore the island to that date. During the winter of 1959–1960, six men overwintered on the ice sheet at Jarl-Joset Station, and the following summer brought a smaller operation of 20 people.[29]

Unsurprisingly given EGIG's French leadership, the European expedition was modeled on the earlier successful EPF expedition to

Eismitte: it relied on weasels and air drops, as well as a division of labor between scientists and technicians to maximize efficiency and prioritize scientific results. Under EPF's management, the international expedition was also able to capitalize on the French–US working relationship that had been built up over the previous two decades. Henri Bader generously provided access to US Army facilities and equipment (including weasels, wanigans, sleds, thousands of liters of fuel, and food rations), and the US Air Force base commander at the 4084th Air Base Group, Col. Byron B. Webb, ensured that EGIG received full support from Sondrestrom Air Base.[30] US support for the cooperative European expedition was rooted in a confidence born of EGIG's leader and team: EGIG "promis[es] very valuable scientific results judging by the caliber of scientific personnel associated with Mr. Victor's expeditions in the past and by the fine record of publications resulting from his expeditions," wrote US glaciologist and former SIPRE director Albert L. Washburn in 1955.[31] Convinced of the expedition's scientific value, the US Army granted EGIG permission to retrieve the American cache near Søndre Strømfjord, which had been left behind by Carl S. Benson's Project Jello team at the end of their traverse in August 1955.[32] Accessing this cache, however, was not simple: by the time the EGIG team arrived at the site in 1959, accompanied by US military officers, the cache was completely buried by snow, "except for the roofs of the wanigans and the tops of the sleds and the weasel pods."[33] Digging out the snow grave was an exercise in frustration, but the results were more than worth the effort: with supplies ranging from radios to medical equipment to leftover frozen foods (including diced beef, butter, green beans, sweet potatoes, peaches, canned peanut butter, and noodles with cheese), the Jello cache represented both a logistical boon and a concrete symbol of US backing for EGIG.

Probing the Ice

Comprising glaciology, geophysics, seismology, meteorology and geodesy, EGIG's scientific program was designed to extend the research conducted by the Wegener, EPF, and Project Jello expeditions at Eismitte right across a horizontal band of the ice sheet. EGIG's scientific planners—France's Paul-Emile Victor and Albert Bauer, Switzerland's Richard Haefeli, German glaciologist and cartographer Richard Finsterwalder, Austrian geophysicist Peter Steinhauser, and Denmark's Børge Fristrup and Helge Larsen—enunciated two main goals: first, to build on measurements taken by previous expeditions in order to confirm and improve the existing scientific understanding of the ice sheet and, second, to pursue climate- and environmental-related questions that had been raised by previous work at Eismitte.[34] Ambitious and forward thinking, EGIG's scientific program would provide tantalizing evidence of atmospheric contamination and climatic change in a decade when these issues were gaining traction in the scientific world.

The EGIG expedition's path traced an asymmetrical cross stretching from Disko Bay in the west to Cecilia Nunatak in the east, and from Jonction northwards to Point Nord—over 1,400 kilometers of snow and ice. Scientific work across this expanse included remeasuring the snow markers initially placed by EPF and later studied by Project Jello, and extending those markers at ten-meter intervals from Eismitte to Greenland's eastern coast. Taken together, the three expeditions spanned a decade of glaciological work, providing "far more accurate data than ever concerning the accumulation, ablation, and the movement of the ice cap," in the assessment of the Greenland Ministry's Eske Brun.[35]

From the very conception of EGIG, Eismitte played a central role in the expedition's scientific program. In the summer of 1955, Johannes Georgi, who had overwintered at Eismitte a quarter century earlier, argued that EGIG should capitalize on the long history of observations at Eismitte by "organiz[ing] observations on a continuing periodic basis in the vicinity of the Central Ice Cap Station."[36] EGIG's scientific committee took Georgi's suggestion to heart: while the French members of the committee were, of course, thrilled to have the opportunity to return to their Station Centrale, commitment to the location ran much deeper. For all countries involved, Eismitte had the aura of a scientifically sacred place, one which hard work over the previous decades had built into the premier scientific location on Greenland's ice sheet. This sentiment redoubled when the French station was again refound in 1959, lying ten meters below the snow surface: it was crushed almost to nothing, the timber beams reduced to half their original height and the prefabricated station walls unrecognizable.[37]

The scientific work undertaken at Eismitte over the four years of the EGIG expedition was widely varied, reflecting the diverse interests of the participating countries. As part of their cross-Greenland traverses, small groups of German scientists conducted geodesic, seismic, and gravimetric studies at Eismitte, viewing the location both as of exceptional scientific value and as a natural place to pause on their route.[38] By elongating existing snow markers and erecting new ones, these studies also contributed to the ongoing effort to understand the movement of the ice sheet and of Eismitte itself. For the Swiss, too, Eismitte formed a focal point for their glaciological and meteorological research. Led by Marcel de Quervain, director of the Swiss Federal Institute for Snow and Avalanche Studies, the Swiss group conducted snow and ice measurements and meteorological observations at Eismitte in order to extend the data series that stretched back to the Wegener expedition. Disappointingly, though, the Swiss team could not locate EPF's ice core drilling channel despite a diligent search, and so their planned remeasurement and resampling of that channel had to be canceled.[39]

But the most important research conducted by EGIG at Eismitte was rooted in questions of climate and environment—questions that were propelled by the changing conception of climate at the time. At the end of the nineteenth century, it was widely agreed that climate was a stable phenomenon that had experienced no significant changes in thousands of years. This view persisted into the early twentieth century, when climate continued to be considered static except on the geological timescale. Historian Matthias Heymann describes climatological research at this point as *classical climatology*, which

> saw its task in the collection and evaluation of climatologic data in order to produce, complete and refine the quantitative description of the climates on earth and provide proper databases for the investigation of the effects of local climates on vegetation, agriculture and human health. It, secondly, attempted to come to an understanding of regional and local climates by investigating geographical and physical influences like mountain ranges, ocean currents, or prevailing winds... Classical climatology maintains the conception of climate that emphasized the stability of climate over time and its variability with respect to geographic location.[40]

Until the 1940s, climate was seen as an ensemble of data about the "normal" weather phenomena (e.g., temperature, wind, precipitation) of a given region, liable only to limited variability. In 1935, for example, the International Meteorological Organization adopted the 30-year period from 1901–1930 as the standard "climatic normal period," underlining the common belief that such a norm was recognized, in effect, by nature itself. This conception permeated the training and work of climatologists at the time. "This view of climate as effectively constant, undergoing at most cyclical changes of no great importance, was taught to most of my generation in school," wrote English meteorologist and climatologist Hubert Horace Lamb, founder of the Climatic Research Unit at the University of East Anglia, in 1964: "It was believed that you only had to average data over a long enough period of time to arrive at a figure to which the climate would always return."[41]

Gradually, however, climate began to shed this stable geographic mantle. In the middle of the twentieth century, the concept of "normal" climates was increasingly questioned and, soon, the idea that a particular set of decades could represent a region's "normal" climate was rejected. Climatologists began to see climate as a dynamic concept, one that could change over short timescales. The emergence of a dynamic view of climate was motivated in large part by the growing recognition of warming trends between the 1890s and the 1940s—trends which contradicted the climatic stability at the heart of classical climatology. "Over the 30 to 50 years from 1900 on, most parts of the world underwent a progressive warming so marked that it could not be overlooked," Lamb explained in a 1964 lecture to the British Association for the Advancement of Science: "Interest in the subject of climatic change was aroused once the considerable warming of our climate in most seasons of the year from the 1890s to the 1930s and '40s became obvious to all."[42] Among those who sounded the alarm on warming were Swedish glaciologist Hans W. Ahlmann, German climatologist Helmut E. Landsberg, and Lamb himself.[43] By the time of the EGIG expedition, the idea of climatic consistency had been dealt such a heavy blow that there was even speculation about an ice-free Arctic Ocean within decades. With this transformation of the concept of climate, Greenland's ice sheet beckoned to EGIG's scientists as a door to understanding past and future climates.[44]

Led by Swiss glaciologists André Renaud of the Helvetic Society for the Natural Sciences and Hans Oeschger of Bern University's Institute of Physics, the six-member EGIG ice core team made its mark on the international research scene by building an improved drill and undertaking core studies across central Greenland. As the emphasis was on analytic techniques and not drilling technologies, most of the EGIG ice cores were relatively short, averaging 20 meters in depth. The apex of the drilling program took place at Eismitte, chosen for its long history of ice core research. There, the EGIG team drilled the deepest core of the entire expedition, reaching down 31 meters into the ice, between June 9 and 13, 1959. The core was carefully cut into 39 pieces, packed in snow for preservation, and shipped to Switzerland for analysis.[45]

EGIG's ice core analysis comprised two main components: tritium and lead studies, both of which hinged on the ice sheet as an undisturbed repository for fallen atmospheric debris. The tritium analysis, undertaken by Oeschger in Bern, was initially intended as another method of dating snow layers to determine annual accumulation. As it turned out, the Swiss tritium analysis also shed light on a little understood climatic phenomenon: the circulation of contaminants and pollutants through the atmosphere. Continuously produced by cosmic radiation, tritium has a half-life of just over 12 years, making its activity measurable along the Swiss Eismitte ice core. This core, however, held a surprise: instead of the expected exponential decrease in tritium concentration with depth due to radioactive decay, the upper layers of the Eismitte core showed higher than expected levels of tritium—a result, Oeschger argued, of tritium deposits from thermonuclear bomb tests.[46] Displaying a clear signature of the first Russian H-bomb tests of summer 1953, the EGIG core demonstrated that ice cores could be used to monitor and track radioactive fallout and contamination. A nuclear physicist by training and an early proponent of anthropogenic climate change, Oeschger went on to found the Division of Climate and Environmental Physics at Bern University in 1963, where he directed work on ice cores until his retirement three decades later. In the same vein, working with the EGIG ice cores in Bern, University of California at San Diego chemist Edward D. Goldberg measured lead-210, a natural radioactive species in atmospheric precipitation with a half-life of 21.4 years, in a polar ice cap for the first time.[47] These studies provided a first step in what would become, in the following decades, a standard method of probing past atmospheric circulation, volcanic activity, fossil fuel burning, and nuclear weapons testing.[48]

Danish physicist Willi Dansgaard performed a third set of analyses on EGIG's Eismitte core—analyses that did not form part of the original plan, but which imparted valuable information about the climatic regime of the ice sheet. Since 1955, when Dansgaard's early work on the analytical potential of oxygen isotope ratios in water was boosted by Samuel Epstein's analysis of the Project Jello ice cores, Dansgaard had continued to develop his ideas and techniques. Together with Norwegian scientist Per Scholander, Dansgaard collected thousands of vials of water from glaciers in Iceland and Greenland, testing them for isotopic content and trapped air bubbles. When he heard about the European expedition to Eismitte, Dansgaard immediately contacted his compatriot and longtime colleague Børge Fristrup and requested access to the ice cores. This access was quickly granted, as EGIG was obliged to accommodate Danish scientists—but, to Dansgaard's dismay, the materials that arrived in Copenhagen were not as rich as he had hoped. A fire at EGIG's laboratory in Paris had destroyed hundreds of kilograms of ice cores: "they all melted," recalls Dansgaard, "but fortunately was possible to save a representative sample from each of the containers."[49] With the rescued core samples and a hefty dose of analytic ingenuity, Dansgaard was able to determine the air temperature at the time of accumulation from the isotopic content of the samples. Stretching back half a century to 1909, the 31-meter deep Eismitte core clearly showed what Dansgaard described as "the improvement of the climate up till the mid forties followed by a cooling."[50] Supported by more than two dozen other ice cores from across central Greenland, this signature of a polar warming in the first half of the twentieth century followed by a cooling on the order of one to three degrees Celsius was "of utmost importance" at a time when this cooling was still poorly understood. "This was an issue full of political dynamite," wrote Dansgaard: the cooling, he worried, would spell an end to the delicate Greenland cod industry—and, indeed, within ten years the cod, which had filled Greenland's waters in the middle of the century, had disappeared.[51] Dansgaard's treatment of EGIG's Eismitte ice core as a climatic indicator proved to be, in the words of his *New York Times* obituary, "a major advance in studying the climate history of the planet, providing evidence that predated other sources of measurement like tree rings, lake sediments and petrified organic matter."[52] Dansgaard went on to perform isotope analyses on the first ice core to reach all the way through Greenland's ice sheet to bedrock, recovered in 1966 at the US military's Camp Century. At 1,387 meters deep, the Camp Century ice core reached back more than 100,000 years into the past, and Dansgaard's analysis provided a complete climatic signature back to the last Ice Age.[53]

An International Expedition on Danish Territory

The most contentious aspect of EGIG, and certainly the aspect that caused the greatest strife for all concerned, was the relationship between the expedition leaders and the host country, Denmark. Whereas Danish authorities had been relatively comfortable with the 1948–1953 EPF expedition to Greenland, they were much more reticent about the EGIG expedition—despite the inclusion of Denmark as a member country. Beginning with Victor's initial 1955 meeting at the Danish embassy in Paris, requests for permission for the international team to travel to Greenland were met with stony silence.[54] The Danes were perturbed by the participation of so many countries and concerned that EGIG's planned coastal studies would infringe on the agendas of Danish scientists. As the only area of Greenland in which Danish glaciologists had historically worked, the coastal zones were a touchy subject, the Danish ambassador to France, Ejnar Wærum, informed EPF's assistant director, Charles Gaston Rouillon, in late 1955, and intellectual sovereignty over this realm was in the country's first interest—a position consistent with the scientific nationalism practiced by Denmark in meteorology, geology, and cartography since World War II.[55] "During our meeting, the ambassador showed a certain impatience, and did not seem to understand that we are pressed," wrote an exasperated Rouillon after a meeting in which Wærum stonewalled the Frenchman's efforts to win support for the expedition.[56] Danish territorial sovereignty was also a concern, wrote Eske Brun to Victor: given the sensitivities in Denmark over the continued US military presence in Greenland, Brun explained, the Danish government was reluctant to allow EGIG to use US airfields for fear that such use would result in "too much publicity around the service that the US Air Force will be rendering you."[57] This reluctance stemmed from twin concerns at the heart of the sovereignty dilemma: perception and precedence. Nils Svenningsen, Permanent Secretary at the Danish Foreign Ministry and former lead negotiator with the German occupiers during the war, was wary about the domestic perception of concrete US support for EGIG: to appease domestic concerns, US activity in Greenland needed to be firmly controlled, not expanded. Granting the European cooperation permission to use the US defense areas, Svenningsen worried, would also create an unwanted precedent, and would make it more difficult for Denmark to turn down similar requests in the future.[58]

Frustrated and increasingly convinced that the expedition might be terminated before it even began, EGIG's leaders decided in early 1956 to put their personal contacts to work. Paul-Emile Victor, who had cultivated a warm relationship with the Greenland Office during his previous work on the island, went to Copenhagen to speak personally to the Permanent Secretary, Eske Brun. He also undertook a letter-writing campaign, addressing, among others, Brun, Helge Larsen, and the Danish ambassador in Paris.[59] Other members of EGIG's leadership, too, traveled to Denmark to plead their case: in early 1956, more than a year after the initial enquiry, Albert Bauer and Richard Haefeli went to speak to the committee responsible for weighing EGIG's request, including geodesists Einar Andersen and Col. Jorgen Helk, geographer Niels Nielsen, and geologists Alfred Rosencrantz and Arne Noe-Nygaard.[60] Emphasizing the intent of the expedition to focus on the interior of the ice sheet, these entreaties aimed to reassure the Danes that EGIG would not trample on their vested coastal interests. Nielsen and Noe-Nygaard, both fervent scientific nationalists, proved the hardest to please. Where official channels failed, this massaging of personal connections succeeded in the nick of time. In the spring of 1956, only weeks before the planned beginning of the expedition, Ambassador Wærum informed EGIG's leaders that permission been granted, subject to the usual restrictions: two Danish representatives, Børge Fristrup and Helge Larsen, had to be allowed to join the expedition committee; a Danish controller would be installed in Greenland (at EGIG's expense) to oversee the expedition; and all scientific results stemming from EGIG had to be published in the Danish periodical *Meddelelser om Grønland*.[61] Denmark also claimed priority for filling the expedition's scientific posts—but in reality, Danish glaciological expertise was limited to the coastal zones and hence this stipulation had little effect on the expedition's scientific plans. The main Danish contribution came from the Geodetic Institute of Denmark, which, under the leadership of Jorgen Helk and supported by French helicopters, worked on Greenland's western coast with the object of producing the first modern topographic map of the area.[62] In the end, Svenningsen even relented on the question of EGIG using US bases in Greenland, writing to Brun that "[we must] not create a precedent for other countries' use of the defense areas for non-defense purposes, but in this case I think, all things considered, we will be best served by taking a favorable attitude to the plan"—a decision taken in large part because the Permanent Secretary of the Foreign Ministry was reluctant to say no to his NATO allies.[63] He did, however, require the French Armée de l'Air to file detailed flight plans before every sortie and to provide the Danish Army Geographical Service with copies of all aerial photographs.

While Victor and his EGIG team were satisfied with the results of their negotiations, the Danes did not enjoy such peace of mind: as the expedition ran its course, Eske Brun—who felt the weight of Greenland on his shoulders—grew increasingly frustrated and, eventually, resentful of EGIG's disregard for their Danish hosts. Brun had been a fixture on Greenland's political scene since he first arrived on the island in 1932, at the age of 28, to take up a temporary position as the governor of northern Greenland. His promotion to a permanent position in 1939 was thrown off course when, the next year, Germany invaded metropolitan Denmark and Greenland was cut off for the duration of World War II. Together with the Danish representative in Washington, Henrik Kauffmann, and the governor of southern Greenland, Aksel Svane, Brun spent the war years "trying to hold together mighty Greenland, which was suddenly cut off from Mother Denmark's supplies" by organizing the transport of essential goods from North America—a dangerous and difficult endeavor which embroiled the three Danes in a struggle for power and authority.[64] The war over, Brun was appointed vice director of the Greenland Office and, two years later, he succeeded Knud Oldendow as Permanent Secretary—and when the Office became the Greenland Ministry (*Grønlandsministeriet*) in 1954, Brun retained his top position. A driving force in Greenland's postwar modernization, Brun is a controversial figure in Denmark, especially in light of his oversight of the forcible dislocation of northwestern Greenland's Uummannaq community in the winter of 1953 to make way for the US Thule Air Base.[65]

As early as 1956, the opening year of the EGIG expedition, Brun had to reprimand Victor for failing to publish the expedition's work in the Danish periodical *Meddelelser om Grønland* as was clearly required by the expedition's permissions.[66] The next year, the French Geodetic Institute photographed and measured nearly two dozen of Greenland's coastal glaciers even though "the French institute hardly had the authority to carry out the work; they had only received permission to work on the inland ice and to photograph access to the inland ice, but not otherwise to work in the coastal zone."[67] Brun was doubly embarrassed by the publication of these maps, since, as noted in an internal meeting, "the French now boast publications and maps of 19 glaciers, when the Danes have carried out only a single glacier measurement."[68] When the Danes were reduced to having to ask the French to send them copies of these photos, sentiment turned against the international cooperation and Brun, together with other Danish scientific and administrative figures, began to speak openly about rejecting any request for a second cooperative European expedition to Greenland.

In 1961, when Victor approached the Danish authorities with another expedition in mind, he found a combative atmosphere. Organized out of Paris and with little in the way of Danish leadership or scientific participation, the dean of Danish glaciology, Børge Fristrup, charged, EGIG was "totally dominated by France" and by "Victor's overwhelming influence."[69] Likewise, by only minimally involving Denmark in the expedition's initial preparations, complained an agitated Niels Nielsen, Victor had created a "thing which Denmark has to accept or which puts Denmark in an awkward position by refusing the program"—and, he continued sternly, "we do not wish to be put into that situation once again."[70] "Denmark must be absolutely free in the future to choose the organizations and countries they wish to collaborate with," emphasized Fristrup.[71] Despite Greenland belonging to Denmark, Eske Brun continued, it was France which got credit for the expedition and which gained a reputation for "doing a good job serving international science."[72] This reputation was built in part through Victor's well-oiled publicity machine, which promoted the expedition—and France's leading role—at public events including film soirées.[73] The schism between Denmark and the other EGIG member countries deepened when the Danes refused to approve Victor's request for a major new expedition, and insisted that a Danish organization would soon replace EPF as the leader of glaciological work in central Greenland—a topic taken up in the epilogue. For now, we continue on Eismitte's path with another US excursion to the middle of the ice sheet.

A Race to Map the World: HIRAN

During World War II, Allied shells fired in the French-German border area landed well past the German line more often than not, since the maps available were riddled with inaccuracies. "The US Army lacked the most vital information needed to correct the maps," wrote John Dille in a popular 1965 article on mapping; that is, "geodetic control points in Germany which could be matched up with control points on the maps of France."[74] Even on friendly territory, World War II-era mapping was far from ideal: established almost entirely by optical methods (e.g., taping distances and measuring angles with theodolites), geodetic control was restricted by line of sight—and, moreover, there was no consistency of references or grids between national maps. For modern war, wrote Floyd W. Hough, first director of the US Army Map Service's Geodetic Division, "an accurate figure of the Earth becomes more and more important...when we consider the use of guided missiles and other long range weapons."[75] In the immediate postwar years, US geodesists estimated that less than three percent of the earth's surface had been surveyed in detail sufficient for modern operations. And soon thereafter, as military and political planners of all stripes adjusted to the new international order, it became increasingly clear that no existing mapping systems or methodologies were capable of adequately handling distances on the continental level. The need for an accurate, worldwide geodetic system to link the continents became acute with the opening of the Cold War and the subsequent development of intercontinental ballistic missiles. "With 5,000-mile missiles about to be added to their arsenal of deterrent weapons," continued Dille, "they [the US military] must know exactly where on earth they might have to aim them. 'Almost exactly' is not good enough."[76] At NATO, planners pointed to a standardized, accurate geodetic reference system able "to keep pace with missile capabilities" as a keystone of collective defense.[77] "On a worldwide scale," a worrying NATO study reported, "we are not sure of the position of North America in relation to the Rurasian continent."[78] Soon, a "race to map the world" emerged between the major Cold War powers.[79]

In the United States, the Air Force spearheaded a project to design a global geodetic system for the accurate targeting of missiles over long distances.[80] Essential to this project was the precise determination of distances between locations on continents separated by vast oceans, described at the time as "the kind geodetic fact which may win or lose a war."[81] One of the systems that came out of this work was known as HIRAN: an aerial surveying technique to establish distances between an organic network of ground locations, described in 1954 as "a geodesist's measuring tape capable of spanning over 500 miles at a single measure."[82] HIRAN's origins date to 1938, when Radio Corporation of America electrical engineer Stuart William Seeley realized that radio transmission signals could be used to measure distances. Seeley had been working on an experimental television system, but quickly recognized that his discovery could be adapted into a blind navigation mechanism for aircraft. The US Army Air Forces agreed, and contracted Seeley to build a blind bombing system in 1941. The resulting system, known as SHORAN (SHOrt-RAnge Navigation), was able to guide bombers to their targets in situations when visual targeting was impossible.[83] In 1942, SHORAN was first installed on military test flights, and in the winter of 1944 Seeley's system was put to work in combat over northern Italy.

At the end of the war, top US surveyor Carl I. Aslakson was tasked with adapting SHORAN's airborne electromagnetic surveying system to measure long distances over the earth's surface.[84] A civil engineer by training, Aslakson had joined the US Coast and Geodetic Survey in 1924, at the age of 27, and served on geodetic and precision measurement missions across the Philippines, Alaska, and the continental United States. With the US entry into World War II, he transferred to the Army Air Forces and conducted surveying operations in the China-Burma-India Theatre and South America before taking on the SHORAN project. By extending geodetic control up to 800 kilometers, Aslakson's work on SHORAN received immediate acclaim. In northern Canada, SHORAN was used to tie the country's remote Arctic archipelago to its central landmass, a key task for exerting sovereignty over vast northern regions.[85] In the United States, the technique was brought to the Aleutian Islands, where incessant fog had impeded hydrographic charting for years. "With Shoran," explained a justifiably proud Seeley, "it was completed in a few months."[86] And oil companies prospecting in the South American jungle and the Gulf of Mexico leapt at SHORAN for position control.

When Aslakson returned to his home base at the US Coast and Geodetic Survey after the war, the Air Force took over SHORAN and again improved its accuracy to within an error range of less than one meter per 200 kilometers.[87] The new system, known as HIRAN (HIgh-precision shoRAN), established distances by trilateration between ground stations and an airborne station.[88] An airplane would fly repeatedly between two ground stations while transmitting pulses to those stations. Measurement of the reflected pulses was then used to calculate the sea-level great circle distance between the ground stations. Crucially, with HIRAN, error did not increase with the distance measured, making it ideally suited for intercontinental mapping.

With its ability to accurately map long distances over remote areas, including ice and water, HIRAN was well placed to forge a geodetic connection between North America and Europe. By linking Canada's Baffin Island to Norway via Greenland and Iceland in what became known as the North Atlantic tie, photomappers wove a geodetic ribbon between the continents. This tie operated with a precision of 12 meters—an improvement of more than 30-fold from the best data available only years before.[89] With similar new mappings from all parts of the globe not cut off by the Iron and Bamboo Curtains, by the early 1960s the US military had gathered enough geodetic data to construct a single global coordinate system for missile targeting.[90] Known as World Geodetic System 1960, or WGS 60, this system—described by historian John Cloud as "arguably one of the most important American intellectual achievements of the Cold War"—provided targeting information for US intercontinental ballistic missile fleets.[91]

Eismitte as a Geodetic Node

The importance of Greenland as a geodetic stepping-stone between the continents is clear. In 1956, the US Air Force established six HIRAN stations on the ice sheet in a rough grid between the 68th and 71st parallels. One of these Greenland stations, designated HIRAN Station 31, was located at Eismitte, just steps south of the buried French station. Eismitte was selected by negotiation between the Air

Force, which was ultimately in charge of HIRAN, and the US Army's Snow, Ice and Permafrost Research Establishment, home to many of the US military's experts in polar ice sheet environments.[92] Just as the Air Force needed SIPRE's expertise to ensure that Greenland's HIRAN stations and personnel could function on the ice sheet, SIPRE saw the proposal for geodetic stations in central Greenland, and especially at Eismitte, as a boon for meteorological work and the long-term study of ice flow. Indeed, HIRAN's Greenland stations were a joint project between the US Air Photographic and Charting Service's geodetic mappers and SIPRE's glaciological experts: working side-by-side, photographic and charting service personnel set up the geodetic equipment while SIPRE took care of logistics and snow markers. After acclimatizing to central Greenland's altitude at a training facility, the four photomappers at each site—bearded from months of relentless work, their faces obscured by giant snow goggles—spent their time on the ice sheet monitoring aircraft signals that, after being reflected back from the ground stations, would be processed into accurate surface distance measurements.[93] To the mappers, whose regular work schedules brought them from coral atolls in the Pacific to the Nigerian rainforest, and from the deep-sea Texas Towers off the US eastern seaboard to the Libyan desert, Eismitte was just another workplace—albeit one where they could spend their downtime skiing and snowshoeing.[94]

At Eismitte, the HIRAN camp boasted a combination of standard, portable Air Force equipment and ice sheet specific constructions: next to the lightweight, prefabricated Quonset huts which the mappers used all over the world stood snow tunnels and walls, all located east of the station's radar mast. Dark against the bright ice sheet, the semicylindrical Quonset huts were fastened down by guide wires that crisscrossed in every direction to keep the huts in place as winds ripped across the island. Long, low white walls of saw-cut snow blocks protected the huts from the worst of the drifting snow, and snow tunnels and trenches allowed the men to travel between the shelters without setting foot outside. The entire camp was dwarfed by the HIRAN antenna, reaching 15 meters into the sky. Through the spring and summer of 1956, contact with the outside world came from high-flying reconnaissance aircraft on geodetic missions and low-flying resupply airdrops that brought food and fuel to the photomappers. Sharing the cooking on a weekly rotation, the men ate standard rations—Vienna sausage, lima beans, pork and gravy, canned bacon and eggs, and some fruit—and melted snow in large pots for drinking and washing. The camp's only color was provided by the Danish and US flags, flapping hard in the wind on low bamboo flagpoles, their edges rough and frayed.[95]

The red and white *Dannebrog* flying at Eismitte, and Danish flags at Greenland's other HIRAN stations, had been requested by US Air Force Col. Robert W. Gates, commander of Task Force HIRAN. A decorated pilot who had earned a citation from President Roosevelt for his flight missions over France during the Normandy Invasion, Gates oversaw two Arctic projects in the mid-1950s: as well as commanding Project Ice Skate, which built a 1,500-meter runway on a floating ice island near the North Pole, he also was responsible HIRAN's operations in Greenland.[96] "It is respectfully requested that the Danish Liaison Officer provide the Task Group Commander with six flags of Denmark," he wrote to the Danish liaison at Sondrestrom Air Base as part of his effort to secure final Danish permission for HIRAN's Greenland stations in March 1956: "These flags would fly side by side with the United States flags, on Danish Sovereign Territory, the Greenland Ice Cap."[97] Gates' note, written just nine days before the HIRAN mission to Greenland was set to begin, also gave Danish authorities the first explicit description of the photomapping project. The HIRAN sites, he explained, "will be utilized to transmit electronic signals to highflying electronic reconnaissance aircraft to accurately measure and map part of the Greenland Ice Cap." The Danes had been aware of the impending project since 1952, when the United States applied for permission to conduct a major airborne geodetic research project over Iceland, Greenland, and northern Canada. Danish authorities were at ease with this project since, as part of a worldwide endeavor directly connected to NATO priorities, it did not infringe on the Danish cartographic agenda. The Danish Foreign Ministry quickly approved HIRAN on condition that the resulting survey data be supplied to Denmark.[98] After two years of work setting up stations in other parts of the North Atlantic, the US provided more details to the Danish Foreign Ministry in 1954, explaining that "we require the establishment of six (6) US Air Force sites generally located between 68° N and 71° 30' N in the central portion of the Greenland ice cap" to complete the North Atlantic tie.[99] The Danes reiterated permission for the project at an extended six-day meeting between Danish and US officials in Copenhagen the following spring. Final Danish clearance for HIRAN was given on June 18, 1956—after the project was well underway.[100]

Scientific Partnerships

The Snow, Ice and Permafrost Research Establishment's scientific interest in the HIRAN station at Eismitte was twofold. First, SIPRE hoped that meteorological measurements taken by HIRAN personnel during their summer work would provide much-needed data about polar whiteouts. A "peculiar meteorological state in which the light from the sky and the light reflected from the surface of the snow are of the same brightness, so that all contours are blotted out," in the words of Børge Fristrup, polar whiteouts are one of the most unpredictable and perilous dangers of the ice sheet.[101] Polar whiteouts posed significant problems for the US military's daily operations in Greenland and were a central focus of US glaciological and meteorological studies on the island through the 1950s and 1960s. Most of the available data, however, came from the vicinity of Thule Air Base in the island's northwestern corner. Even as weather modification experiments were being devised and implemented in the early-to-mid 1950s, there was little pertinent data from the center of the ice sheet. These experiments, which ran for ten years and employed more than a dozen Army scientists, aimed to clear whiteouts by seeding menacing clouds with dry ice delivered by manned aircraft, drones, rockets, and mortar shells.[102]

The Air Force's proposal for six HIRAN stations in central Greenland, including one at Eismitte, struck SIPRE's Administrator James E. Gillis as a valuable opportunity for gaining additional whiteout data.[103] In early 1956, two months before Greenland's HIRAN stations became operational, Gillis wrote to the Air Force inquiring as to "the possibility of obtaining [your] weather reports to support...our studies on white-out and visibility in the Arctic."[104] Gillis specifically requested that HIRAN personnel provide details about ice crystals and fog during whiteouts to supplement SIPRE's data from northern Greenland. "Observations at the HIRAN sites where the sun angle and air temperature differ from [our sites] in North Greenland might help us in our studies on white-out and visibility in the Arctic," he emphasized. Cognizant of the value of such meteorological data for air operations in polar environments as well as the need for a working cooperation with SIPRE, the Air Force responded positively to the request. "APCS [Air Photographic and Charting Service] personnel at each site have been given a brief weather observer indoctrination course," answered US Air Force Lt. Col. R.G. Bounds, adding in a slightly worrying note that "reasonably accurate observations are anticipated."[105] In addition to their geodetic duties, the HIRAN personnel at Eismitte took meteorological observations every six hours between May and August 1956. In the autumn of that year, the data was relayed to SIPRE's chief scientist, Henri Bader, and incorporated into whiteout and weather modification studies.

SIPRE's second scientific interest in HIRAN's Eismitte station was the accuracy with which the location of the station's radar antenna was known—a result of it being one node in a geodetic network designed specifically for accuracy. This knowledge was a boon to the long-standing interest in the dynamics of the ice sheet, including the potential climatic implications of ice melt. From as far back as Wegener's expedition, Eismitte had been a base point for understanding the movement and mass balance of the ice sheet: how did Greenland's ice move and flow through the seasons, was the island's ice sheet as a whole growing, stable or shrinking, and how would ice sheet melt affect the world's oceans and low-lying areas? These questions provided a concrete link between the scientific teams that visited Eismitte over the years: the snow markers planted by the French EPF team at Eismitte and along a trail from Eismitte to Quervains Havn between 1949 and 1951 were located, remeasured and extended by Carl S. Benson's Project Jello team in 1955, and then by the EGIG team several years later. At Greenland's HIRAN sites, SIPRE personnel erected telescoping aluminum markers projecting 11 meters above the snow surface and catalogued them with respect to the stations' radar masts.[106] Based on existing knowledge of snow accumulation in central Greenland, SIPRE expected the markers to remain visible until 1963—an estimate that would turn out to be badly off.[107] "Observations repeated periodically will permit the study of correlation between the movements and variations of the inland ice and long term climatic variations," explained Swiss glaciologist and EGIG's chief scientist Richard Haefeli.[108] Thanks to HIRAN Station 31, Eismitte's location was known more accurately than ever before, and any movement of the station's markers over time could be closely tracked.

In a display of commitment to this line of research, the US Army mounted an expensive mission to find and lengthen the snow markers at all six of central Greenland's HIRAN sites in 1963, the last year in which the original markers were still expected to protrude above the snow.[109] Led by glaciologist Steven J. Mock and geologist Donald L. Alford of the Cold Regions Research and Engineering Laboratory (CRREL) and supported by a 12-person aircrew from the US Air Force's 17th Tactical Airlift Squadron, the team took off from Sondrestrom Air Base in May 1963 in search of the sites.[110] As members of CRREL's Materials Research Branch, Mock and Alford had experience conducting glaciological, geodetic, and seismic studies in north Greenland and stateside, and approached their mission confidently. Faced with the enormous, desolate and monochromatic ice sheet, however, the 17th Squadron navigators "were unanimous in their estimate of the chances of successfully finding isolated aluminum poles several hundred miles out on the ice cap: Zero!," recalled Mock.[111] But, thanks to years of well-maintained markers at Eismitte, HIRAN's Station 31 leapt out on the radar. Using Eismitte's strong radar target as a point of departure and blessed by "an inordinate amount of luck," Mock and Alford were able to locate three more of Greenland's HIRAN stations in the following days, but the others remained hidden under the snow.[112] Standing at Eismitte on May 11, 1963, looking at the snow markers that had been installed by the French and later extended by the Americans and then by the European cooperation, Mock and Alford were hopeful that continued measurements in the years to come would ultimately lead to a better understanding of the dynamics of Greenland's ice sheet.

At that point, however, Mock and Alford's hope was thin: despite more than a decade of measurements and remeasurements, the question of Eismitte's movement was desperately confused. The 1950s and early 1960s saw eight measurements of Eismitte's snow markers.[113] The two measurements conducted by EPF showed that Eismitte had moved 100 meters between 1950 and 1951, but this result was thought to be unreliable because of the short time period between the measurements. Upon arriving at the old French station in 1955, George Wallerstein, Project Jello's navigator, calculated that Eismitte had moved nearly 800 meters to the south in the four years since the French departure—a result that Børge Fristrup called "rather astonishing."[114] Puzzled, EPF's Paul-Emile Victor hired Louis Tschaen, a professor at the National School of Engineering in Strasbourg, to assess the situation. Without ever setting foot in Greenland, Tschaen reexamined all previous measurement data and found that Eismitte was moving 170 meters a year to the southwest. Even the HIRAN antenna at Eismitte couldn't help unravel the growing mystery since the US Air Force closely guarded its newly acquired, strategically valuable geodetic data.[115] In 1959, EGIG added to the confusion by conducting two more measurements, one astronomic and one geodetic, and declaring that Eismitte had hardly moved at all from its 1950 location. The only results from more than a decade of work were frustration and uncertainty: was the central ice sheet moving at a rapid rate towards the south (or southwest), or was Eismitte stationary to within the margin of error of the measurement instruments? This uncertainty underlined two problems: first, the limitations of conventional surveying and geodetic techniques and, second, the inherent difficulties of measuring a single location.

On Greenland's ice sheet, the lack of stationary reference points hampered tellurometers and theodolites, making it difficult to determine absolute ice velocities.[116] "The use of classical methods is hopeless," bemoaned French glaciologist Albert Bauer, who himself had helped with several of the calculations, in 1968: "the precision of photogrammetric measurements from aerial surveys is much superior."[117] As the 1960s turned into the 1970s, Bauer's vision became reality when the expansion of radar and satellite technologies transformed ice sheet movement measurements. New technologies—among them, airborne synthetic-aperture radar, satellite tracking, laser and radar altimetry, and automated time-lapse digital photography—circumvented the problems associated with conventional instruments, increased the accuracy and the scope of ice sheet measurements, and allowed arrays of disparate points to be studied with minimal surface setup.[118] The introduction of giant deformation polyhedrons, too, allowed angles and distances to be carefully tracked, removing the uncertainties inherent in single-location studies. In a typical example, the summers of 1980 and 1981 saw a joint US-Danish-Swiss team use the US Navy Navigational Satellite System to monitor 22 stations in three polyhedral clusters across the southern portion of the ice sheet, capturing data from approximately 13,000 satellite passes.[119] In this new world of ice sheet measurement, Eismitte did not stand out: rather, these measurement projects viewed the island from a bird's eye perspective, working at multiple sites spread across its entirety. With the rise of remote sensing, the ice sheet could increasingly be studied without being visited, again lessening the importance of physical locations. This broadening of perspective provides a window into Eismitte's declining importance in the latter third of the twentieth century.

Fading into the Ice: Eismitte in a New Era

Early in the twentieth century, when Alfred Wegener was planning his return to Greenland, the very center of that great island stood out as one of the last unexplored places in the Arctic world. Both getting to Eismitte and surviving there through the polar winter were fraught with difficulties—indeed, it was precisely these difficulties which led to Wegener's death in the early winter of 1930. By applying new technologies—among them, airdrops, polar tractors, and helicopters—the postwar French, American and cooperative European expeditions transformed access to Eismitte.[120] Postwar quarters at Eismitte were also a far cry from the German undersnow cave: the scientists enjoyed prefabricated shelters, fresh produce, music, and regular communication with families back home. Still, with the exception of the HIRAN station, discussed below, the participants of the major postwar expeditions to Eismitte all traveled overland to the center of the ice sheet: landing aircraft on the unprepared ice sheet, not to mention taking off again, was seen as too risky.[121]

How things had changed by the time Steven J. Mock and Donald L. Alford arrived at Eismitte in 1963! The Air Force's 17th Tactical Airlift Squadron was able to land their plane directly on the ice sheet and taxi to within 15 meters of Eismitte, obviating the need for overland transport and airdrops. The ability to land aircraft on unprepared snow surfaces was the result of years of work by the US military and by private companies.[122] The first landing on unprepared snow on Greenland's ice sheet, however, was anything but planned: in 1928, during an attempt to fly their custom-built blue-and-yellow single engine Stinson Detroiter from Rockford, Illinois, to Stockholm, Sweden, US flight pioneers Bert (Fish) Hassell and Parker (Shorty) Cramer got blown off course over southern Greenland. Running low on fuel and nearly blinded by the sun reflecting off the snow, the pair brought their plane down on the open ice sheet 16 kilometers inland—and then spent a harrowing fortnight navigating the marginal zone until they were rescued by a scientific party from the University of Michigan.[123] Even as Hassell and Cramer enjoyed a triumphant return to the United States, where they were celebrated at a ticker-tape parade in New York and greeted by President Calvin Coolidge and President-elect Herbert Hoover in Washington, unprepared ice sheet landings weren't seriously considered as a nonemergency measure until after World War II.

On June 8, 1947, the Commander General of the US Air Force's Atlantic Division, Major General William H. Turner, and the Air Force's Assistant Chief of Staff for Operations, Major General Earle E. Partridge, shared an air inspection flight over Greenland. Gazing out over the immense white expanse, they wondered aloud about the possibility of air landings on the ice sheet—but, upon enquiring, were told that there was no available data.[124] Not satisfied, Turner took it upon himself to pursue the idea. Drawing on the US Navy's Operation Highjump, which had brought ski-wheeled C-47s to the Antarctic the previous year, Turner launched a project to assess the possibility "of flying personnel and equipment directly into the central portion of the [ice] Cap."[125] Within two months of Turner's spring flight, the Air Force's foremost Arctic rescue expert, Lt. Col. Emil Beaudry, and seven men from the Atlantic Division were operating 160 kilometers inland from Søndre Strømfjord under the banner of Project Snowman.

After three weeks of reconnaissance, Beaudry's men successfully landed a ski-wheeled C-47 on an unprepared section of Greenland's ice sheet on August 21, 1947—but when they attempted to take off again, "the skis were found to be frozen solid to the ice [and the] second take-off attempt succeeded only after a considerable run through the soft snow."[126] With five more unprepared landings and takeoffs on the ice sheet in August and September, Project Snowman provided proof of concept—but also concluded that modifications to the planes were necessary before takeoff from soft snow could be considered practicable.[127] The following year, Beaudry put Project Snowman's work to the test when he successfully landed his C-47 on the ice sheet in hazardous conditions to rescue 12 men who had been stranded for three weeks after losing control of their plane during a whiteout and crashing.[128] And on May 3, 1952, ace fighter pilot and Air Force Lt. Col. William Benedict guided his ski-wheeled C-47 to the first successful American landing at the North Pole, allowing the Air Force to claim victory over the Navy in the US interservice North Pole aviation race.[129] In the mid-1950s, C-47s also resupplied two radar stations on the ice sheet near Thule and brought photomapping personnel to central Greenland's HIRAN sites, relying on jet assisted takeoff rocket systems to provide the power necessary to lift off in the soft snow.[130]

In the lead up to the IGY of 1957–1958, the US Navy was eager for large-capacity, long-range ski-equipped aircraft to support its work in the Antarctic. The C-47s used for unprepared snow landings in the early-to-mid part of that decade were deemed too small for the jobs that lay ahead on the southern continent. American aerospace giant Lockheed took on the task of building a bigger ice-sheet-capable plane. In 1957, Lockheed installed skis on a four-engine turboprop C-130 and began landing tests on frozen lakes in Minnesota. By fitting giant six-meter long skis weighing one ton each around conventional landing gear and adding extra underwing fuel tanks and jet rockets, Lockheed delivered a behemoth of an airplane that could take off from a proper runway and land on an unprepared snow surface, or vice versa, and which could generate the enormous thrust needed to lift off from soft snow. Boasting a payload more than seven times greater than that of the C-47, the C-130 lived up to its nickname of *Hercules*. Within five years of Lockheed's first tests, the Navy bought four of the ski-equipped C-130s for its Antarctic operations and the Air Force took possession of 12 of the new planes. In 1963, one of these Air Force planes ferried Mock and Alford directly to Eismitte.

Mock and Alford were awed by the Air Force's ability to land on unprepared snow surfaces. Unable to hide their enthusiasm even in formal reports to the Army Corps of Engineers, the glaciologist and geologist touted the C-130 as "an ideal support airplane" with "the versatility and range... to establish and supply field parties on any part of the ice cap [and to] eliminate much of the lost time associated with surface transport and supply."[131] Their view was widely shared, and from that year on scientists and military personnel traveled regularly by air throughout Greenland.[132] In the Antarctic, too, "aviation has progressed to the point where inland flying is routine," wrote US Antarctic Projects Office staff historian Henry M. Dater in 1963: beginning in that year, he continued, "field parties and stations in any part of the continent are feasible."[133] The C-130's versatility, cargo capacity, and long-range transformed work on Greenland's ice sheet: with the planes, any point of the island's interior could be reached and made livable regardless of whether it had previously been visited by an overland convoy. The risks, certainly, were high: with low clouds, ice fog, and strong winds constantly reforming the surface of the ice sheet, unprepared snow landings were uncertain at best. Over Greenland, pilots fly "against a grim, inanimate enemy—the ice, the winds, the cold," wrote newspaper reporter Bill Duncan after a 1964 visit to the C-130 aircraft *in situ*: "It is as though nature is angry because man has set foot in a forbidden white paradise."[134] The official flight handbook for the ski-wheeled C-130 ordered pilots to disregard normal safety factors and to treat ski operations as a calculated risk: deep snow tricked instruments, the ice sheet's high altitude reduced engine performance, and takeoff from soft snow, even with jet assistance, was harrowing. These risks were laid bare by the crash of a US C-130 in southern Greenland in 1968: upon landing hard in the snow, the main landing gear and one ski separated and were thrown 200 meters away from the damaged body of the airplane, which came to rest heavily on its portside wing, the propeller broken off and driven into the snow.[135] Still, even with these risks, ski-equipped aircraft normalized access to the ice sheet, opening it to work that before would have been inconceivable.

As accessing the ice sheet became a more routine proposition, the primacy of Eismitte melted away. No longer did the central ice sheet location stand out as it had before: the great overland journeys made by the Wegener, EPF, and Project Jello expeditions to the very middle of the ice sheet—journeys that stood out as much for their transportation exploits as for their scientific results—were no longer necessary, replaced by direct air access. "Scientific exploration of Greenland, like life in Greenland in general, has completely changed," Fritz Loewe, formerly of the Wegener expedition, said in 1969: "Research workers fly to and from Greenland in a few hours. They land from helicopters at any desired place. [And] modern methods of research and transport make it possible to solve old questions which could not even be approached at the time when they appeared."[136] Further, with the rise of remote sensing technologies, networks, or arrays, of locations grew in importance. Hints of these transformations are, in fact, visible through the four decades of scientific expeditions to Eismitte considered here: for the Wegener and EPF expeditions, Eismitte was a singular point of interest, the focus of intense year-round scientific research, a destination as much as an objective. For Project Jello, Eismitte stood out as the most important location on a much longer trek, singled out by its history of scientific work. And for both the US Air Force HIRAN project and the cooperative European EGIG expedition, Eismitte was part of a larger endeavor, still distinguished by its history but no longer as dominant: for HIRAN, Eismitte was one of six nodes in a grid of geodetic stations in central Greenland, while for EGIG, Eismitte was an important stop for an expedition that crisscrossed over the central and southern portions of the island.

Eismitte's decline is also linked to the establishment of two major semipermanent research stations on the ice sheet. Known as DYE-2 and DYE-3, these stations formed part of the Distant Early Warning (DEW) Line, a 6,000-kilometer long necklace of radar stations stretching from Alaska across northern Canada and Greenland to Iceland, designed to detect Soviet bombers approaching over the pole. As the northern edge of an electronic grid system ultimately controlled from the North American Aerospace Defense Command (NORAD) hub in Colorado, the DEW Line was a bastion of North American continental security through the Cold War.[137] The first foundations of the DEW Line were laid in 1954 across Canada and Alaska, and the system went online in mid-1957. Under order from the Joint Chiefs of Staff to extend the line eastwards, Greenland's DYE-2 and DYE-3 stations were built two years later, in 1959–1960. The location of these ice sheet stations, on the southern part of the island, was chosen to follow the curve of the nearly 60 DEW Line stations already in operation.[138]

Greenland's ice sheet presented a special challenge for the DEW Line's construction crew. Because of radar and communications requirements, the stations needed to be above ground, at the mercy of winds and blowing snow, bearing the full brunt of Greenland's ferocious winters. The Northeast Air Command, tasked with the project, was so overwhelmed by the complexities of ice sheet work that they advised the US Air Force to drop the ice sheet stations entirely—but

Air Force Headquarters overruled this advice.[139] The stations that rose from Greenland's interior in 1959–1960 were giant prefabricated black boxes topped and sided with plastic radomes, compound white bubbles encasing the all-important radar antenna. All parts and supplies for the radar stations, from bolts and screws to huge steel beams and trusses, were flown to the ice sheet on ski-equipped C-130s from Sondrestrom Air Base. Held nine meters above the snow surface on eight chunky legs, the stations squatted over the ice sheet like spaceships come to rest in a vast, frozen white ocean. This construction allowed snow to blow under the stations and prevented drifts from piling up—and, thanks to hydraulic jacks housed in the legs, the entire stations could be raised to compensate for accumulated snowfall over the years. At six stories tall and weighing 2,400 tons each, the DYE stations were the heaviest buildings ever erected on ice. The DEW Line personnel at each station enjoyed well-heated living areas with paneled walls, elegant hanging light fixtures, posters and paintings, as well as leatherbacked chairs and wooden desks. In the dining room, red tablecloths were topped with real china plates, good food, wine, and, in honor of the host country, Danish aquavit.[140]

As well as linking into North America's early warning grid, Greenland's DYE stations also provided a natural base for glaciological research on the ice sheet. With enormous logistical capacity, including vehicle shelters, fuel storage, living quarters, proper sewage, sports and recreation facilities, diesel generators, and a well-maintained 3,000-meter long snow runway, they were in effect fully equipped, semipermanent research stations. Scientific investigations of the ice sheet, Børge Fristrup announced in 1961, "can be carried out there under much better conditions...and at much less expense" than at less well-equipped locations, including Eismitte.[141] Through the 1960s and 1970s, US and Danish scientific groups rotated in and out of the DYE stations on a regular basis, conducting snow drift, ice core, and geodetic studies.[142] For the new generation of Danish glaciologists, led by ice core pioneer Willi Dansgaard, DYE-3 became their go-to "test station for new glaciological techniques, because it offered workshops and accommodation in the field."[143] Even in cases when other ice sheet locations would have been more valuable from a scientific point of view, the DYE stations took precedence for logistical and financial reasons. During the Greenland Ice Sheet Project (GISP), a joint Danish-US-Swiss ice core project which ran from 1971–1981, for example, "although available [data] pointed to the optimum site location for the first deep drilling to be in north-central Greenland," wrote US glaciologist Chester C. Langway, "the financial restrictions forced the selection of the logistically convenient Dye-3 location."[144] Using US and Danish drilling technologies, Langway—together with Willi Dansgaard and Hans Oeschger, both of whom had cut their teeth on ice cores from Eismitte—drilled through the ice sheet to bedrock at DYE-3, recovering a 2,037-meter long ice core in 1981, the deepest from Greenland and the second deepest in the world at that time.[145]

In this new era, Eismitte faded into the ice, no longer occupying a prominent place in the scientific imagination. As more knowledge was gained about Greenland's interior, the mystery of Eismitte lessened; as access to the ice sheet opened, the act of getting to the middle of the island lost its allure; and as remote sensing technologies came to the fore, physical access to the ice sheet diminished in importance. Finally, with the establishment of fully equipped, semipermanent stations on the ice sheet, the gravity of scientific research on the island shifted. But still for those men who traveled to Eismitte in the first two-thirds of the twentieth century, the location retained symbolic value and emotional attachment. Reminiscing during a public lecture in 1969, nearly 40 years after spending the winter of 1930–1931 at the center of Greenland's ice sheet, Fritz Loewe painted Eismitte as the keystone to "a new period of systemic scientific research" in the Arctic—and, he added proudly, "by far the coldest station on earth."[146] Loewe carried his memories of Eismitte more heavily than most: a tall man, he walked with an awkward gait, his body pitched forwards, a permanent reminder of the toes he lost to severe frostbite during that first overwinter in the middle of the island. Loewe closed his address by capturing the heart of the changes that had swept over Greenland in the four decades since he and his German colleagues first lay their eyes on Eismitte: in this new era, he said in a voice spilling over with melancholy and wonder, "One flight across the ice sheet gives 100 times more information than could be gathered in one season 40 years ago."

Epilogue

A Conspicuous Absence

Perhaps the most puzzling aspect of the story of Eismitte is the near absence of Danes at the center of Greenland's ice sheet. Through four decades of scientific activity at Eismitte, the national interests of other countries were clearly on display, from the Weimar Republic's desire to make its mark on the polar regions following World War I to France's decision to develop the know-how to defend its territorial claim in the Antarctic to the United States' continental security concerns during the Cold War. But what about Denmark? Despite Greenland being a part of Denmark, Danish activity at Eismitte was limited to administrators and observers dispatched by the Foreign Ministry and the Greenland Ministry to keep an eye on foreign scientists.[1] Even during Expédition Glaciologique Internationale au Groënland, where Denmark was one of five European partners, no Danes took part in scientific work on the ice sheet. To understand this seemingly peculiar situation, we need to step back to the early nineteenth century.[2]

Until 1814, Greenland—along with Iceland and the Faroe Islands—formed part of the Norwegian possessions of King Frederik VI of Denmark, who ruled over the union of Denmark-Norway. With the defeat of Denmark-Norway at the end of the Napoleonic Wars, the union of the two kingdoms dissolved. While mainland Norway was ceded to the King of Sweden, Denmark maintained sovereignty over Norway's overseas possessions, including Greenland.[3] But the Norwegians never took their eye off the great island. "In the final decades of the 19th century," writes historian Christopher Jacob Ries, "the Norwegians moved their activities closer to the East Greenland sea-ice"—and, foreshadowing the scientific nationalism that would sweep over Greenland after World War II, "Denmark responded by increasing expeditions to the region... conducted by Navy personnel traveling by boat or ship combining assertions of national sovereignty with marine biological investigations and topographical surveys along the coast."[4] After regaining independence from Sweden in 1905, Norway made a claim to Greenland based on the island's pre-1814 status as a Norwegian possession. In 1931, supported by a circle of Greenland activists in Oslo, Norwegian whaler Hallvard Devold began a private occupation of the region of eastern Greenland he called *Eirik Raudes Land*, or Erik the Red's Land, a stretch of coast between 71° 30' North and 75° 40' North.[5] The Norwegian government soon announced support for the occupation on the basis that the area was a *terra nullius*. Denmark brought Norway to the International Court of Justice in The Hague and, two years later, the dispute was finally resolved when the International Court granted Denmark full sovereignty over Greenland.[6]

The International Court's ruling in favor of Denmark was based in part on the long tradition of Danish scientific activity in Greenland. In the years leading up to that court decision, as tensions over Greenland were rising, Denmark redoubled its late nineteenth-century scientific program with the intention of "securing total scientific and political dominance over the last remains of its colonial empire."[7] In 1930, in parallel with swelling Norwegian sympathy for the Greenland cause, the Danish government commissioned geologist and Arctic explorer Lauge Koch to undertake a large-scale scientific expedition to eastern Greenland. And when Norway laid claim to Erik the Red's Land the following year, Denmark responded with the massive Three Year Expedition to East Greenland 1931–1934, which saw more than 100 scientists descend on Greenland under Koch's leadership. "Armed with new field technologies, such as permanent wintering stations, airplanes, and radios," Ries writes, Koch's party of scientists "inaugurated a new era in Arctic research under Danish leadership."[8] With the International Court's decision, intellectual sovereignty was effectively transformed into juridical sovereignty.

The twin concepts of intellectual (or epistemological) sovereignty and juridical sovereignty, as well as the closely related environmental authority, are currently attracting attention in and being worked out by the polar history community.[9] It is beyond the scope of this book to elaborate on them in detail, except to note that the pursuit of power through knowledge has been central to Denmark's approach to Greenland for over a century. Geographic, cartographic, and other scientific knowledge were mechanisms for asserting control over a space: knowing a space was a means of justifying authority over that space in the absence of more traditional symbols of sovereignty, and, later, creating new knowledge about a space was a means of actively bolstering that authority in face of real and perceived challenges. Science was, in effect, co-opted as a means of state control over territory.[10]

In the early postwar years, when it became clear that the United States military was in no hurry to leave Greenland, Denmark again turned to science as a means of exercising sovereignty over the island. A national program of scientific research in Greenland, historian Peder Roberts has written, was "a powerful means of asserting relevance within an increasingly superpower-dominated world."[11] Following the war, the Danish government looked to meteorology, cartography, and geology—all sciences with strong connections to the performance of sovereignty—as tools to regain momentum over vast tracts of land that had been politically separated from occupied metropolitan Denmark through the war. Despite having few resources and fewer trained personnel, Denmark took control of US weather stations in Greenland immediately after the war, operating them first through the Greenland Office and then, after 1954, through the Danish Meteorological Institute.[12] In recognition of the tight connection between mapmaking, or territorial knowledge, and sovereignty, the *Grønlands Geologiske Undersøgelser* (Greenland Geological Survey) was founded in 1946 to produce geological maps of western Greenland. In the same year, Denmark's *Geodætisk Institut* (Geodetic Institute) resumed its mapping activities on the southern part of the island.[13] And in 1947, the Danish government backed Eigil Knuth and Ebbe Munck's expedition to northeast Greenland's Peary Land, "deciding that the relative absence of Danish cartography there—and its proximity to Siberia—made performing sovereignty imperative."[14] Cartography was particularly important to the Danish sovereignty enterprise given both the country's long history of cartographic activity in Greenland and the silent threat to this established dominance from the United States. The NATO-backed 1951 Defense of Greenland Agreement stated that if Denmark was unable to furnish "topographic, hydrographic, coast and geodetic surveys and aerial photographs, etc. of Greenland" as requested by the United States, the United States could undertake to procure them themselves—a situation which Denmark was eager to avoid, equating it with a declared lack of control over the island.[15] "Greenland as a part of Denmark," Eske Brun emphasized, "is a thousand-year-old historical fact."[16]

This active exertion of scientific authority over Greenland was complemented by a parallel policy of scientific nationalism. Immediately following World War II, Swedish glaciologist Hans W. Ahlmann proposed a pan-Scandinavian expedition to Greenland to pursue meteorological and glaciological studies. His plans were quickly dashed when Denmark declined to participate—a decision which Roberts describes as part of "a growing conviction that resources must be concentrated on *national research* precisely in order to prevent...the United States [from running] the show in Greenland."[17] In the late 1940s and 1950s, Paul-Emile Victor found himself reliving Ahlmann's experience: when Victor proposed a joint French-Danish expedition to Eismitte in 1948, he was rebuffed by Danish authorities who placed precedence on purely national contributions to scientific work in Greenland. The following year, Victor offered his fully equipped Station Centrale overwinter station at Eismitte to Danish researchers, asking in return that Denmark maintain the station and that French scientists be granted continued permission to visit and conduct investigations at Eismitte. But a central ice sheet station didn't fit in the Danish scientific plan for Greenland. Denmark rejected the offer and Station Centrale was left to be buried by the snow.[18] And during Expédition Glaciologique Internationale au Groënland, Danish authorities were highly protective of Danish intellectual interests in Greenland's coasts, nearly refusing permission for the expedition as a whole because it posed a threat to those interests.

Denmark's conspicuous absence at Eismitte is a product of both these vested interests and the country's lack of glaciological experience on the ice sheet. Historically, Danish scientists had concentrated their efforts along Greenland's coasts. The 1870s saw a rash of Danish geological expeditions to western Greenland, and the following decade brought geographical exploration of eastern Greenland's then largely unknown coast—an endeavor pushed by rising Norwegian activity in eastern Greenland. In that era, Danish glaciology, too, focused on the coastal regions, with Count Carl Moltke's 1894 study of glaciers at Nordre Sermilik Fjord and Ludvig Mylius-Erichsen's 1906–1908 *Danmark* Expedition.[19] In the post–World War II years, under Børge Fristrup's leadership, Danish glaciologists maintained this coastal agenda. Trained in glaciology at Stockholm University's Kebnekaise Station, perched on the side of Sweden's tallest mountain, Fristrup spent every summer (and some winters) from the late 1940s on in Greenland, leading research and amassing an encyclopedic knowledge of the island's coastal glaciers. Soon after the war, he led investigations around Academy Glacier and Navy Cliff at the opening of Independence Fjord as part of Knuth and Munck's broader scientific (and primarily geographical) exploration of northern Greenland. And during the International Geophysical Year of 1957–1958, Denmark's glaciological contribution centered on studies of four glaciers outside the ice sheet proper. "[A]ll the stations were local glaciers outside the Ice Cap, and only the measurements at the Hurlburt Gletcher in the Thule district have some relation to Ice Cap studies," wrote Fristrup.[20] This trend persisted during the cooperative European expedition, when Danish scientists restricted themselves to coastal work despite having the right to join any part of the expedition.[21] Through these decades, Danish glaciology involved measuring the areas and volumes of coastal glaciers, studying their oscillations and relation to climate fluctuations, and taking aerial photographs to determine changes in the glaciers' shape and size by means of comparison to past descriptions, drawings, and maps.

As the 1960s opened, Danish glaciologists shed their focus on Greenland's coasts and began to look into the ice sheet itself. This newfound interest was driven by the bitter aftermath of the EGIG expedition: having authorized the cooperative European expedition under pressure, Danish political and scientific authorities soon found themselves outraged with the expedition's disregard for its Danish hosts and embarrassed that Denmark received so little recognition for a successful expedition on Danish territory (cf. chapter 4). To prevent a similar situation in the future, the Greenland Ministry's Permanent Secretary, Eske Brun, proposed that Denmark take control of glaciological research in central Greenland. "Denmark is ready to fulfill its part in the future," he announced in 1961: to lead work in ice sheet glaciology and to put Danish scientists at the head of work situated on Danish territorial soil (or, rather, snow).[22] Described by his foreign counterparts as "outspoken [and] too blunt for good diplomatic relations," Brun had a long history of protectionism towards the island for which he was responsible: he vehemently (and successfully) opposed the creation of a new US defense area in northeastern Greenland in the early 1950s and, under his eye, fraternization between US servicemen and indigenous Greenlanders was banned.[23] "The work on the Ice Cap was going on on Danish ground just as Danish as the ground in Copenhagen," thundered Brun at a meeting of Danish and foreign glaciologists in Copenhagen: "Denmark has the sovereignty over the Ice Cap [and] we want a part in the leadership in the explorations that is comparable to the fact that Greenland is Danish."[24] It was essential, Fristrup had advised Brun prior to the meeting, that future ice sheet research be organized out of Copenhagen, not Paris or Washington or any other foreign capital.[25] These sentiments were bluntly echoed by Niels Nielsen, the renowned geographer and vice-rector of the University of Copenhagen who was still seething from the EGIG expedition: Denmark must "be completely free to make up our mind what ought to be done and to choose our way of organizing and to decide with whom we would cooperate," he told the slightly shocked assembled group of foreign scientists.[26] As he listened to these words, the significance of the new Danish attitude dawned on Jean Vaugelade, assistant director of Expéditions Polaires Françaises. "The elements of the problem are not the same as before; the political problems are not the same and neither are the operational," Vaugelade said: now, he continued, capturing the crux of the issue, "Denmark insists that Greenland is Danish."[27]

Coupled with this desire to take charge and prevent a repeat of the EGIG experience was a growing recognition of the importance of Greenland's ice sheet for environmental and climatic issues. From the time of the Wegener expedition's 1930–1931 overwinter at Eismitte, two main themes had gradually taken hold in the scientific and public imaginations. First, inspired by Ernst Sorge's demonstration of a link between snow density and climate, digging and drilling deep into the ice proved a rich method of understanding past climates and the circulation of contaminants and pollutants through the atmosphere. With ice cores, to travel down into the ice sheet is to travel back in time: quite literally, as Danish researcher Willi Dansgaard did in the 1960s, to touch snow which had fallen during the Eemian interglacial more than 100,000 years ago. Second, together with growing information about polar warming, ice sheet thickness measurements enabled quantitative calculations of the environmental and human consequences of melt from Greenland's ice sheet. With vivid descriptions of low-lying populated areas being submerged by rising waters, these calculations grabbed headlines and, more soberly, raised far-reaching but elusive questions about the political and social consequences of ice melt.[28] In Denmark, recognition of Greenland's unique ability to shed light on these environmental and climatic questions took hold. "The Greenland Ice Cap is the largest mass of land ice outside the Antarctic ice sheet," wrote Brun in a 1961 letter outlining Denmark's growing interest in the ice sheet: and, he continued, "therefore studies of the behavior of this ice cap [are of] special importance."[29] And in Willi Dansgaard, Brun had the perfect vehicle for his ambitions: Dansgaard's touch for extracting paleoclimatic knowledge from ice cores had established him as one of three leading experts in ice core research in the world.[30]

Still, glaciological studies on the ice sheet, including ice core research, were dependent on logistics and infrastructure in Greenland's interior—an area in which Denmark was comparatively weak. Even Dansgaard had never set foot at Eismitte: his ice core samples from central Greenland were extracted by EGIG and flown to Paris before finally being sent to Dansgaard's Copenhagen laboratory. On the political side, during an official visit to Greenland in July 1957, the Canadian Ambassador to Denmark, Herbert F. Feaver, had to rely on US helicopters for travel over the island when Danish authorities could not provide the necessary transport.[31] And when a Danish citizen, Borgs Petersen, was struck with acute appendicitis at Station Nord, a Danish-run weather station in northernmost Greenland, in January 1963, he was evacuated not by Danish personnel but by a US Air Force C-130 dispatched from Sondrestrom Air Base. Petersen was flown to the US hospital at Thule Air Base, where, as a press release boasted, "a successful emergency operation...put Mr. Petersen in an 'excellent condition' status."[32] These were far from isolated incidents: US planes and personnel routinely assisted with medical, political and other needs across the island.[33]

With no significant ice sheet facilities and little experience running glaciological operations on the ice sheet proper, the chances of Denmark taking charge of ice sheet research in Greenland seemed slim. Fristrup was realistic about the situation, emphasizing to Brun in 1961 that "if Denmark can or will not become active on the ice sheet, I think we should give [foreign scientists] permission for the surveys and then let them implement them without excessive interference."[34] This was, however, far from Fristrup's preferred option: "On the other hand, if Denmark does not actively engage in the work now, it will be difficult to intervene later," he continued, "and it seems to me unfortunate if the Greenland ice sheet should be the workplace of numerous foreign enterprises without real Danish influence." But with the establishment of three semipermanent US research stations on the ice sheet in the late 1950s and early 1960s, Fristrup's worries were alleviated and Denmark's logistical problems were solved. Capitalizing on the open scientific access to US sites in Greenland afforded to them through bilateral agreements, Danish glaciologists made extensive use of the research facilities at the DYE-2 and DYE-3 Distant Early Warning Line stations and at the shorter-lived Camp Century. Under Dansgaard's leadership, a new generation of Danish glaciologists based at his University of Copenhagen laboratory was at the helm of a string of high-profile results, from the construction of a complete climate signature back to the last Ice Age based on the Camp Century deep ice core to the identification of abrupt glacial climatic leaps in a deep ice core from DYE-3 to the development of the microchip-controlled "Rolls Royce drill," a lightweight and transportable ice core drill so-named for its speed and reliability.[35] The mode of research that underpinned this work was one rooted in logistical, technical, scientific, and financial cooperation with other countries. From the mid-1960s on, spurred by the high costs of ice core research, Danish glaciologists collaborated with researchers from more than ten partner nations.[36] Born of both practicality and an increasingly confident Danish glaciological community, this mode contrasted sharply with the scientific nationalism practiced by Denmark in the immediate postwar period.[37]

Upon Brun's resignation from his long-held post as Permanent Secretary of the Greenland Ministry in 1964—an act prompted by his deep disappointment with the persisting pay inequality between Greenlandic and Danish officials in Greenland—the King of Greenland's determination to secure Danish control over ice sheet research had arguably laid the groundwork for a significant Danish contribution to ice sheet science.[38] After nearly a century of focusing on Greenland's coastal glaciers, in the 1960s Danish glaciologists turned to the ice sheet proper: a shift of attention and resources stimulated by growing dissatisfaction with Denmark's historical lack of influence over scientific work on the ice sheet and by the realization that Greenland's interior was key to understanding increasingly pressing environmental and climatic questions. And central to this realization were the four decades of research at Eismitte considered in this book—research that, from the early days of the first overwinter in the middle of the ice sheet in 1930–1931, shed light on the shape, movement, and melt of Greenland's ice, on the circulation of contaminants through the atmosphere, and on the earth's climatic past.

Notes

Major Archival Collections Consulted
1. Consulted in Copenhagen and at Erhvervsarkivet (the Business Archives) in Aarhus.

Introduction The Edge of the World, the End of the World
* Paul-Emile Victor et al., *Groënland: 1948–1949* (Paris: Arthaud, 1951), 28. Except where indicated, all translations have been done by the author.
1. Ibid.
2. Michael Spender and Therkel Mathiassen, "Alfred Wegener's Greenland Expeditions 1929 and 1930–31: Review," *The Geographical Journal* 84 (1934): 515.
3. Snowfall, or accumulation, in Greenland is notoriously variable, and hence averages such as the one provided here need to be understood carefully. See John Maurer, *Local-Scale Snow Accumulation Variability on the Greenland Ice Sheet from Ground-Penetrating Radar* (MA Thesis, University of Colorado at Boulder, 2006). Technically, the process of turning snow into glacial ice proceeds from snow to névé to firn to glacial ice.
4. James R. Ryan and Simon Naylor, "Exploration and the 20th Century," in *New Spaces of Exploration: Geographies of Discovery in the 20th Century*, ed. Simon Naylor and James R. Ryan (London: I.B. Tauris, 2010), 11; Roger D. Launius, "Toward the Poles: A Historiography of Scientific Exploration During the International Polar Years and the International Geophysical Year," in *Globalizing Polar Science: Reconsidering the International Polar and Geophysical Years*, ed. James R. Fleming and Roger D. Launius (New York: Palgrave Macmillan, 2010), 53.
5. Felix Driver, "Modern Explorers," in *New Spaces of Exploration: Geographies of Discovery in the 20th Century*, ed. Simon Naylor and James R. Ryan (London: I.B. Tauris, 2010), 245.
6. See, for example, Mark Carey, "The History of Ice: How Glaciers Became an Endangered Species," *Environmental History* 12 (2007); Eric G. Wilson, *The Spiritual History of Ice: Romanticism, Science, and the Imagination* (New York: Palgrave Macmillan, 2003); Barry H.

Lopez, *Arctic Dreams: Imagination and Desire in a Northern Landscape* (New York: Charles Scribner & Sons, 1986); Francis Spufford, *I May Be Some Time: Ice and the English Imagination* (New York: St. Martin's Press, 1997).

7. Launius, "Toward the Poles: A Historiography of Scientific Exploration During the International Polar Years and the International Geophysical Year," 57.
8. Trevor H. Levere, *Science and the Canadian Arctic: A Century of Exploration, 1818–1918* (Cambridge: Cambridge University Press, 1993), 425. Much of this work centered on mapping, that is, delineating land-sea boundaries and identifying islands and other geographic features. See Urban Wråkberg, "The Politics of Naming: Contested Observations and the Shaping of Geographical Knowledge," in *Narrating the Arctic: A Cultural History of Nordic Scientific Practices*, ed. Michael Bravo and Sverker Sörlin (Canton, MA: Science History Publications, 2002).
9. Fridtjof Nansen, *Nord i Tåkeheimen. Utforskningen av Jordens Nordlige Strøk i Tidligere Tider* (Oslo: Kristiania, 1911).
10. Spender and Mathiassen, "Alfred Wegener's Greenland Expeditions 1929 and 1930–31: Review," 515.
11. Particularly interesting works in these domains include Henrika Kuklick and Robert E. Kohler, "Introduction: Science in the Field," *Osiris* 11 (1996); Jeremy Vetter, ed., *Knowing Global Environments: New Historical Perspectives on the Field Sciences* (New Brunswick, NJ: Rutgers University Press, 2010); Kristian Hvidtfelt Nielsen and Christopher Jacob Ries, eds., *Scientists and Scholars in the Field: Studies in the History of Fieldwork and Expeditions* (Aarhus: Aarhus University Press, 2012); Stuart McCook, "'It May Be Truth, But It Is Not Evidence': Paul De Chaillu and the Legitimation of Evidence in the Field Sciences," *Osiris* 11 (1996).
12. The leading work here is David N. Livingstone, *Putting Science in Its Place: Geographies of Scientific Knowledge* (Chicago, IL: University of Chicago Press, 2003). Also see Richard C. Powell, "Geographies of Science: Histories, Localities, Practices, Futures," *Progress in Human Geography* 31, no. 3 (2007); Simon Naylor, "The Field, the Museum and the Lecture Hall: The Spaces of Natural History in Victorian Cornwall," *Transactions of the Institute of British Geographers* 27 (2002).
13. Ronald E. Doel, "Constituting the Postwar Earth Sciences: The Military's Influence on the Environmental Sciences in the USA after 1945," *Social Studies of Science* 33 (2003).
14. Raf de Bont, "Between the Laboratory and the Deep Blue Sea: Space Issues in the Marine Stations of Naples and Wimereux," *Social Studies of Science* 39 (2009): 199. Also see Dag Avango's work on industrial and resource heritage sites: for example, Louwrens Hacquebord and Dag Avango, "Settlements in an Arctic Resource Frontier Region," *Arctic Anthropology* 46 (2009).

15. Richard Burkhardt, *Patterns of Behavior: Konrad Lorenz, Niko Tinbergen, and the Founding of Ethology* (Chicago, IL: University of Chicago Press, 2005).
16. Jonathan Jones, "Greenland's Ice Sheet Melt: A Sensational Picture of a Blunt Fact," *The Guardian*, Friday, July 27, 2012.
17. With the dissolution of the union between the Danish and Norwegian crowns in 1814, the Treaty of Kiel placed Norway's overseas possessions—including Greenland, Iceland, and the Faroe Islands—under the control of the Danish monarch. In 1905, following regaining independence from Sweden, Norway again made a claim to Greenland. The dispute came to a head in July 1931, when the Norwegian government declared support for a private Norwegian occupation of eastern Greenland. Denmark placed Greenland's fate in the hands of the International Court of Justice in The Hague. In its 1933 ruling, weighing "the declaration of Norwegian occupation of 10 July 1931, its legality, its validity" against "Danish title to the sovereignty of Greenland resulting from peaceful and continuous exercise of state authority," the court ruled in favor of Denmark ("Statut Juridique du Groënland Oriental, Cour Permanente de Justice Internationale, 26ième Session, 5 avril 1933" (Leydon: A.W. Sijthoff's Publishing Co., 1933)).
18. 'Metropolitan Denmark' refers to the major islands of Sjælland and Fyn, the Jylland Peninsula, minor surrounding islands, and the Baltic Sea island of Bornholm. It excludes the Faroe Islands and Greenland.
19. For the defense of Greenland during World War II, see William H. Hobbs, "The Defense of Greenland," *Annals of the Association of American Geographers* 31, no. 2 (1941); Hans W. Weigert, "Iceland, Greenland and the United States," *Foreign Affairs* 23, no. 1 (1944); Nancy Fogelson, "Greenland: Strategic Base on a Northern Defense Line," *Journal of Military History* 53, no. 1 (1989); Finn Løkkegaard, *Det Danske Gesandtskab in Washington 1940–1942* (Copenhagen: Gyldendal, 1968). For sovereignty concerns in the postwar era, see Eric S. Einhorn, *National Security and Domestic Politics in Post-War Denmark: Some Principal Issues, 1945–1961* (Odense, Denmark: Odense University Press, 1975); Eric S. Einhorn, "The Reluctant Ally: Danish Security Policy 1945–49," *Journal of Contemporary History* 10, no. 3 (1975); Poul Villaume and Thorsten Borring Olesen, *I Blokopdelingens Tegn, 1945–1972* (Copenhagen: Gyldendaal, 2005); Shelagh D. Grant, *Polar Imperative: A History of Arctic Sovereignty in North America* (Vancouver: Douglas & McIntyre, 2010).
20. Mark Solovey, *Shaky Foundations: The Politics-Patronage-Social Science Nexus in Cold War America* (New Brunswick, NJ: Rutgers University Press, 2013).
21. Ryan and Naylor, "Exploration and the 20th Century," 11. For classic examples of classification in the context of exploration, see Vilhjalmur

Stefansson, *The Friendly Arctic: The Story of Five Years in Polar Regions* (New York: Macmillan Co., 1921); Laurence P. Kirwan, *A History of Polar Exploration* (New York: W. W. Norton & Co, 1960).
22. Wråkberg, "The Politics of Naming: Contested Observations and the Shaping of Geographical Knowledge," 159.
23. Stephen Bocking, "A Disciplined Geography: Aviation, Science and the Cold War in Northern Canada," *Technology and Culture* 50 (2009): 273.
24. Peder Roberts, *The European Antarctic: Science and Strategy in Scandinavia and the British Empire* (New York: Palgrave Macmillan, 2011), 3.
25. Beau Riffenburgh, *The Myth of the Explorer: The Press, Sensationalism, and Geographical Discovery* (London: Belhaven, 1993), 2. On this, also see Elizabeth Baigent, "'Deeds Not Words'? Life Writing and Early 20th Century British Polar Exploration," in *New Spaces of Exploration: Geographies of Discovery in the 20th Century*, ed. Simon Naylor and James R. Ryan (London: I.B. Tauris, 2010).
26. Michael F. Robinson, *The Coldest Crucible: Arctic Exploration and American Culture* (Chicago, IL: University of Chicago Press, 2006). The quote is from Robert Marc Friedman, "Review of Michael F. Robinson's 'The Coldest Crucible: Arctic Exploration and American Culture'," *Isis* 99, no. 3 (2008): 641.
27. Roberts, *The European Antarctic: Science and Strategy in Scandinavia and the British Empire*, 6.
28. See, for example, Børge Fristrup, *The Greenland Ice Cap*, trans. David Stoner (Copenhagen: Rhodos, 1966); Louis Rey, *Groënland: Univers de Cristal* (Paris: Flammarion, 1974).
29. For a broader look at science in Greenland during the Cold War, see Matthias Heymann and Ronald E. Doel, eds., *Exploring Greenland: Science and Technology in Cold War Settings* (forthcoming).

1 A Land Apart

1. In 1911, he changed his name to Jens Arnold Diderich Jensen Bildsøe.
2. Jens Arnold Diderich Jensen, *J.A.D. Jensens Indberetning om den af Ham Ledede Expedition i 1878* (Copenhagen: Meddelelser om Grønland, 1890), 64–65.
3. Ibid., 62.
4. Laurence P. Kirwan, *A History of Polar Exploration* (New York: W. W. Norton & Co, 1960), 191.
5. Hinrich Rink, "The Recent Danish Explorations in Greenland and Their Significance as to Arctic Science in General," *Proceedings of the American Philosophical Society* 22, no. 120 (1885): 280.
6. Fridtjof Nansen, "Journey Across the Inland Ice of Greenland from East to West," *Proceedings of the Royal Geographic Society and Monthly Record of Geography* 11, no. 8 (1889): 469.

7. See, for example, Hugh J. Lee, "Peary's Transections of North Greenland, 1892–1895," *Proceedings of the American Philosophical Society* 82 (1940): 922.
8. Robert E. Peary, "Journeys in North Greenland," *The Geographical Journal* 11, no. 3 (1898): 214.
9. Rink, "The Recent Danish Explorations in Greenland and Their Significance as to Arctic Science in General," 281.
10. Elisha Kent Kane, *Arctic Explorations* (Philadelphia, PA: Childs & Peterson, 1856).
11. Peary, "Journeys in North Greenland," 215.
12. Paul-Emile Victor et al., *Groënland: 1948–1949* (Paris: Arthaud, 1951), 28.
13. Peary, "Journeys in North Greenland," 228.
14. Paars relocated missionary Hans Egede's earlier *Haabets Koloni* (Hope Colony), originally founded in 1721, from Kangeq Island to nearby Godthaab, on the mainland.
15. Nansen, "Journey Across the Inland Ice of Greenland from East to West," 469. With the personal union of Denmark and Norway from 1524 until 1814, Frederik IV was king of both Denmark and Norway, but is commonly referred to as Frederik IV of Denmark. For the early history of Greenland, see Finn Gad, *Grønlands Historie I Indtil 1700* (Copenhagen: Nyt Nordisk Forlag Arnold Busck, 1978); Finn Gad, *Grønlands Historie II 1700–1782* (Copenhagen: Nyt Nordisk Forlag Arnold Busck, 1969).
16. Paars's account appears in Louis Bobe, *Diplomatarium Groenlandicum 1492–1814* (Copenhagen: Meddelelser om Grønland, 1936), 186–89.
17. Quoted in Fridtjof Nansen, "Journey on the Inland Ice," *Journal of the American Geographical Society of New York* 23 (1891): 174.
18. The two old Norse settlements, Østerbygd (Eastern Settlement) and Vesterbygd (Western Settlement), were in fact both situated on Greenland's west coast, with Østerbygd to the south of Vesterbygd. In Dalager's time, and for a long time after, it was thought that Østerbygd was on the island's eastern coast.
19. Quoted in Rud Kjems, *Horisonter af Is: Erobringen af den Grønlandske Indlandsis* (Copenhagen: GEC Gads Forlag, 1981), 22–25.
20. Robert Brown, "Obituary: Dr Hendrik Rink [sic]," *The Geographical Journal* 3 (1894): 65.
21. For example, Hinrich Rink, "Om Isens Udbredning og Bevægelse over Nordgrønlands Fastland Samt om Isfjældenes Oprindelse," *Tidsskrift for Populære Fremstillinger af Naturvidenskaben* (1853).
22. For a more complete list, see Gunnar Jensen, "One Hundred Years of Crossings of Greenland's Inland Ice," *American Alpine Journal* 32 (1990). A detailed look at early explorations of the ice sheet appears in Kjems, *Horisonter af Is: Erobringen af den Grønlandske Indlandsis*; Børge Fristrup, *The Greenland Ice Cap*, trans. David Stoner (Copenhagen: Rhodos, 1966).

23. Nordenskiöld himself traveled 117 kilometers inland, and his Saami companions continued alone a further 113 kilometers before turning back, but the exact distance is disputed.
24. This claim is thought to be exaggerated; see, for example, Nansen, "Journey on the Inland Ice."
25. Rink, "The Recent Danish Explorations in Greenland and Their Significance as to Arctic Science in General," 286.
26. Nansen, "Journey Across the Inland Ice of Greenland from East to West," 470. The crossing is described in Roland Huntford, *Nansen: The Explorer as Hero* (Trowbridge, UK: Duckworth, 1999); Fridtjof Nansen, *Paa Ski over Grønland. En Skildring af den Norske Grønlands-Ekspedition 1888–89* (Kristiania: Aschehoug, 1890).
27. Huntford, *Nansen: The Explorer as Hero*, 114.
28. Nansen, "Journey Across the Inland Ice of Greenland from East to West," 479.
29. Alfred Wegener and Johan Peter Koch, *Durch die Weiße Wüste; Die Dänische Forschungsreise Quer Durch Nordgrönland 1912–1913* (Berlin: Julius Springer, 1919); Johan Peter Koch and Alfred Wegener, *Wissenschaftliche Ergebnisse der Dänischen Expedition nach Dronning Louises-Land und Quer Über das Inlandeis von Nordgrönland 1912–1913* (Copenhagen: Meddelelser om Grønland, 1930).
30. "Captain Koch's Crossing of Greenland," *Bulletin of the American Geographical Society* 46 (1914).
31. This section is by no means meant to provide a comprehensive history of Greenland as a strategic space, but rather an overview with emphasis on the middle of the twentieth century. In addition to the sources suggested throughout this section, also see John J. Teal, "Greenland and the World Around," *Foreign Affairs* (October 1952).
32. For Greenland's pre- and early population history, see, for example, Bjarne Grønnow and John Pind, eds., *The Paleo-Eskimo Cultures of Greenland: New Perspectives in Greenlandic Archaeology* (Copenhagen: Danish Polar Center, 1996); Hans Christian Gulløv, ed. *Grønlands Forhistorie* (Copenhagen: Gyldendal, 2005); Kirsten A. Seaver, *The Frozen Echo: Greenland and the Exploration of North America, ca. A.D. 1000–1500* (Stanford, CA: Stanford University Press, 1996).
33. The National Museum of Denmark hosts important collections on early Greenlandic cultures and settlements. Ongoing archeological work includes the recent discovery of a Paleo-Eskimo site in Inglefield Land, studied by a joint US-Greenland group; see John Darwent et al., "Archaeological Survey of Eastern Inglefield Land, Northwest Greenland," *Arctic Anthropology* 44 (2007).
34. For Norse, or Viking, activity in Greenland, the catalogue from the Smithsonian Institution's 2000 Viking exhibition is highly recommended: William W. Fitzhugh and Elisabeth I. Ward, eds., *Vikings: The North Atlantic Saga* (Washington, DC: Smithsonian Institution Press, 2000).

35. "The Saga of Erik the Red," English translation of the original 'Eiríks Saga Rauða' by John Sephton, 1880, Chapter 2.
36. The academic and popular literatures on the demise of the Norse in Greenland are vast. For two well-written but differing viewpoints, see Jared Diamond, *Collapse: How Societies Choose to Fail or Succeed* (New York: Penguin Books, 2005); Seaver, *The Frozen Echo: Greenland and the Exploration of North America, ca. A.D. 1000–1500*.
37. For the early cartography of Greenland, see William H. Hobbs, "Zeno and the Cartography of Greenland," *Imago Mundi* 6 (1949).
38. Kirsten A. Seaver, "Review of Narrating the Arctic: A Cultural History of Nordic Scientific Practices," *Arctic* 56 (2003): 306.
39. Nancy Fogelson, "Greenland: Strategic Base on a Northern Defense Line," *Journal of Military History* 53, no. 1 (1989); John Edwards Caswell, *Arctic Frontiers: United States Explorations in the Far North* (Norman, OK: University of Oklahoma Press, 1956); Michael F. Robinson, *The Coldest Crucible: Arctic Exploration and American Culture* (Chicago, IL: University of Chicago Press, 2006).
40. "Convention Between the United States and Denmark, 39 Stat. 1706, Treaty Series 629," signed at New York on August 4, 1916. As Shelagh Grant notes, at that time protection of the Panama Canal (for which the Danish West Indies were pertinent) was seen as much more important than presence in the Arctic regions. See Shelagh D. Grant, *Polar Imperative: A History of Arctic Sovereignty in North America* (Vancouver: Douglas & McIntyre, 2010), 248.
41. Paolo Coletta, ed. *United States Navy and Marine Corps Bases Overseas* (Westport, CT: Greenwood Press, 1985), 132. For Greenland during and after World War II, see Clive Archer, "The United States Defense Areas in Greenland," *Cooperation and Conflict* 23 (1988); Warren F. Kimball, *The Juggler: Franklin Roosevelt as Wartime Statesman* (Princeton: Princeton University Press, 1991); Niels Amstrup, "Grønland i det Amerikansk-Danske Forhold, 1945–1948," in *Studier i Dansk Udenrigspolitik: Tilegnet Erling Bjøl*, ed. Niels Amstrup and Ib Faurby (Aarhus: Politica, 1978); Bo Lidegaard, *I Kongens Navn: Henrik Kauffman i Dansk Diplomati, 1919–1958* (Copenhagen: Samleren, 1996); "DUPI Vol. 1, Grønland Under den Kolde Krig: Dansk og Amerikansk Sikkerhedspolitik, 1945–68" (Copenhagen: Dansk Udenrigspolitisk Institut, 1997); Nikolaj Petersen, "Negotiating the 1951 Greenland Defense Agreement: Theoretical and Empirical Analyses," *Scandinavian Political Studies* 21 (1998); Nikolaj Petersen, "SAC at Thule: Greenland in the US Polar Strategy," *Journal of Cold War Studies* 13 (2011); Poul Villaume, *Allieret Med Forbehold: Danmark, NATO og den Kolde Krig, Et Studie i Dansk Sikkerhedspolitik, 1949–1961* (Copenhagen: Eirene, 1995); Jørgen Taagholt and Jens Claus Hansen, *Den Nye Sikkerhed: Grønland i et Sikkerhedspolitisk Perspektiv* (Denmark: Atlantsammenslutningen, 1999); Poul Villaume and Thorsten Borring Olesen, *I Blokopdelingens Tegn, 1945–1972* (Copenhagen: Gyldendaal, 2005).

42. Balchen's story is told in his fast-paced memoirs: Bernt Balchen, Corey Ford, and Oliver LaFarge, *War Below Zero: The Battle for Greenland* (New York: Houghton Mifflin Co., 1944).
43. "Denmark-United States Agreement Relating to the Defense of Greenland (United States Executive Agreement Series, No. 204)," signed in Washington DC on April 9, 1941.
44. Villaume and Olesen, *I Blokopdelingens Tegn, 1945–1972*, 52.
45. For analysis, see Eric S. Einhorn, *National Security and Domestic Politics in Post-War Denmark: Some Principal Issues, 1945–1961* (Odense, Denmark: Odense University Press, 1975); Erik Reske-Nielsen and Erik Kragh, *Atlantpagten og Danmark 1949–1972* (Copenhagen: Atlantsammenslutningen, 1972).
46. "Defense of Greenland 1951, Agreement Between the United States and the Kingdom of Denmark, April 27, 1951" (American Foreign Policy 1950–1955, Basic Documents, Department of State Publication 6446, General Foreign Policy Series 117, Washington DC).
47. Matthias Heymann et al., "Exploring Greenland: Science and Technology in Cold War Settings," *Scientia Canadensis* 33 (2010). The Soviet attitude towards Greenland, and specifically the Soviet response to the buildup of US presence on the island, is harder to assess. In the early 1950s, as Denmark's decision to uncouple itself from its traditional policy of nonalignment was still echoing, the Soviet press launched a vitriolic press campaign against Denmark's membership in NATO. At a much higher level, Marshall N. A. Bulganin, chair of the USSR council of ministers, wrote to Hans C. Hansen, the Danish prime minister and foreign minister, in 1957, stating that "a vast part of Danish territory, Greenland, has already been converted into a military base of the United States and is to all intents and purposes outside Danish control" (N. A. Bulganin, "Letter to Danish Prime Minister and Minister of Foreign Affairs H.C. Hansen" (NATO, RDC/156/57)). "As recent events in the Near East [i.e., Suez] show, the big powers belonging to NATO can unleash military operations even without informing about it the small countries who are their allies under the North Atlantic pact," the letter continued: "The Soviet Government does not wish at all to exaggerate the picture, but allow me to tell you without mincing words that Denmark, by allowing herself to be inveigled in the war preparations of some powers is subjecting herself to a very serious and unjustified risk." If NATO or the United States were to attack, Bulganin emphasized, the Soviets would "take immediate measures for striking a crushing blow at the aggressor and at the entire network of his support points and bases set up for attack on the USSR"—a thinly veiled threat of a Soviet counter-attack on Greenland and perhaps on metropolitan Denmark itself. For Denmark, the Marshall concluded, "the granting of bases to a foreign state [is] tantamount to suicide in case atomic war breaks

out." The Danish Prime Minister, acutely aware of the sovereignty implications of US presence in Greenland and its potential impact on domestic politics, was blunt in his response: "it is important for me to stress that I can definitely refute your assertion that this island should be virtually outside Denmark's control," he wrote to Bulganin in April 1957, referring to the limitations on US presence imposed by the 1951 Defense of Greenland Agreement (H. C. Hansen, "The Danish Prime Minister's Reply to Marshal Bulganin's Letter of 28 March 1957, written on 26 April 1957" (NATO, RDC/177/57)).
48. Frederic S. Ross and Paul E. Ancker, "Thule Air Base," *Tidsskriftet Grønland* 9–10 (1977): 270.
49. North American continental defense is a vast topic. In order to understand its development and extent, see Christopher J. Bright, *Continental Defense in the Eisenhower Era: Nuclear Antiaircraft Arms and the Cold War* (New York: Palgrave Macmillan, 2010); Sanjay Chaturvedi, *The Polar Regions: A Political Geography* (Chichester, UK: John Wiley & Sons, 1996); Allan A. Needell, *Science, Cold War and the American State: Lloyd V. Berkner and the Balance of Professional Ideals* (Amsterdam: Harwood, 2000).
50. Much has been written on Wegener's final expedition. This chapter focuses on what Eismitte as a place meant to the expedition, and on the scientific investigations undertaken at Eismitte—investigations which laid the foundation for the location's importance for decades to come. Among the many works on Wegener and his expedition, the following are recommended: Volker Jacobshagen, ed., *Alfred Wegener 1880–1930, Leben und Werk, Ausstellung Anlässlich der 100. Wiederkehr Seines Geburtsjahres* (Berlin: Dietrich Reimer Verlag, 1980); Roger McCoy, *Ending in Ice: The Revolutionary Idea and Tragic Expedition of Alfred Wegener* (Oxford: Oxford University Press, 2006); Elsie Wegener and Fritz Loewe, eds., *Greenland Journey, The Story of Wegener's German Expedition to Greenland in 1930 to 1931 as Told by Members of the Expedition and the Leader's Diary* (London: Blackie & Sons, 1939); Ulrich Wutzke, *Durch die Weisse Wüste: Leben und Leistungen des Grönlandforschers und Entdeckers der Kontinentaldrift Alfred Wegener* (Gotha: Perthes, 1997).
51. David Thomas Murphy, *German Exploration of the Polar World: A History, 1870–1940* (Lincoln, NE: University of Nebraska Press, 2002), 107. This section on Germany and the polar sphere draws on Murphy's book. Also see Cornelia Lüdecke, "Approaching the Southern Hemisphere: The German Pathway in the 19th Century," in *Globalizing Polar Science: Reconsidering the International Polar and Geophysical Years*, ed. Roger D. Launius, James Rodger Fleming, and David H. DeVorkin (New York: Palgrave Macmillan, 2010).
52. Murphy, *German Exploration of the Polar World: A History, 1870–1940*, 108.

53. Quoted in Ibid., 134.
54. Quoted in Jacobshagen, *Alfred Wegener 1880–1930, Leben und Werk, Ausstellung Anlässlich der 100. Wiederkehr Seines Geburtsjahres*, 28.
55. "Review of Mid-Ice: The Story of the Wegener Expedition to Greenland," *The Geographical Journal* 85, no. 5 (1935): 478.
56. Murphy, *German Exploration of the Polar World: A History, 1870–1940*, 107.
57. Founded in 1920 by leading members of the Prussian Academy of Sciences (including Schmidt-Ott, Max Planck and Fritz Haber), the Emergency Association of German Science aimed to maintain and support the German scientific community in a time of economic hardship. It was renamed the *Deutsche Gemeinschaft zur Erhaltung und Förderung der Forschung* (German Association for the Support and Advancement of Scientific Research) in 1929.
58. Murphy, *German Exploration of the Polar World: A History, 1870–1940*, 132. For the funding of the Wegener expedition, also see G. Stäblein, "Alfred Wegener, From Research in Greenland to Plate Tectonics," *GeoJournal* 7, no. 4 (1983): 365.
59. Hans W. Ahlmann, "Review of Scientific Results of the German Alfred Wegener Greenland Expedition 1929 and 1930–31," *Geografiska Annaler* 23 (1941): 134.
60. First quote: Ursula B. Martin, "Review of Alfred Wegener: The Father of Continental Drift by Martin Schwarzbach, Carla Love," *Isis* 78, no. 2 (1987): 324. Second quote: Murphy, *German Exploration of the Polar World: A History, 1870–1940*, 152.
61. Georg C. Amdrup, *Danmark-Ekspeditionen til Grønlands Nordøstkyst 1906–1908 Under Ledelse af L. Mylius-Erichsen* (Copenhagen: Meddelelser om Grønland, 1913). This tragedy was due in part to inaccuracies in Robert E. Peary's survey of northern Greenland, which showed a channel cutting off Greenland's northeastern corner (known as Peary's Channel) where none existed. At the time of the *Danmark* Expedition, Peary's Channel was shown on the standard map of Greenland published in Denmark. See "The Non-Existence of Peary Channel," *Geographical Review* 1, no. 6 (1916).
62. Quote: Alfred Wegener, "Tagebücher, June 1906-August 1908," p. 93 (Deutsches Museum Archive Nl001/005, available online at http://www.Environmentandsociety.org/Exhibitions/Wegener-Diaries/Overview). Wegener's diary gives a vivid account of his thoughts on this expedition.
63. Stäblein, "Alfred Wegener, from Research in Greenland to Plate Tectonics," 361.
64. Alfred Wegener, *Die Entstehung der Kontinente und Ozeane* (Braunschweig: F. Vieweg, 1915).
65. For the gradual acceptance of continental drift, see Imre J. Demhardt, "Alfred Wegener's Hypothesis on Continental Drift and Its Discussion in Petermanns Geographische Mitteilungen, 1912–1942,"

Polarforschung 75 (2005); Reinhard A. Krause, "Alfred Wegener, Geowissenschaftler aus Leidenschaft: Eine Reflexion Anlässlich des 125. Geburtstages des Schöpfers der Kontinentalverschiebungstheorie," *Deutsches Schiffahrtsarkiv* 28 (2005); Naomi Oreskes, *The Rejection of Continental Drift: Theory and Method in American Earth Science* (New York: Oxford University Press, 1999); Martin Schwarzbach, *Alfred Wegener und die Drift der Kontinente* (Stuttgart: Wissenschaftliche Verlagsgesellschaft, 1986).

66. Ahlmann, "Review of Scientific Results of the German Alfred Wegener Greenland Expedition 1929 and 1930–31," 134.
67. Kirwan, *A History of Polar Exploration*, 322.
68. Wegener's diary quoted in Wegener and Loewe, eds., *Greenland Journey, The Story of Wegener's German Expedition to Greenland in 1930 to 1931 as Told by Members of the Expedition and the Leader's Diary*, 79.
69. In the same years as the Wegener expedition, two other expeditions also overwintered on the ice sheet: the British Arctic Air Route Expedition of 1930–1931 (led by Gino Watkins) and the Greenland Expeditions of the University of Michigan of 1926–1931 (led by William H. Hobbs). Unlike the Wegener expedition, these two expeditions stayed close to the coasts. The British expedition is noteworthy for its overwinter, which saw a single man (Augustine Courtauld) ride out five months of winter utterly alone in his small station—a feat of mental courage as much as physical prowess.
70. See, for example, Knud Rasmussen, "Professor Alfred Wegener in Memoriam," *Geografisk Tidsskrift* 34, no. 2 (1931).
71. "Wegeners Grønlands Expedition," *Geografisk Tidsskrift* 32, no. 1 (1929); Murphy, *German Exploration of the Polar World: A History, 1870–1940*, 136.
72. The Greenland Office, or *Grønlands Styrelse*, was the chief administrative body in Greenland from 1925–1950.
73. Johannes Georgi, *Mid-Ice: The Story of the Wegener Exhibition to Greenland*, trans. F.H. Lyon (New York: EP Dutton & Co, 1935), 19; Rasmussen, "Professor Alfred Wegener in Memoriam."
74. Ahlmann, "Review of Scientific Results of the German Alfred Wegener Greenland Expedition 1929 and 1930–31."
75. Japanese meteorologist Wasaburo Ooishi had encountered jet streams a few years earlier, but his work was not well known outside Japan.
76. Quoted in Jutta Voss, "Alfred Wegeners Weg als Polarforscher," in *125 Jahre Deutsche Polarforschung*, ed. Alfred Wegener Institut (Bremerhaven: Alfred-Wegener-Institut für Polar- und Meeresforschung, 1994), 88.
77. Alfred Wegener, "Tagebücher, April 1930-September 1930," pp. 133, 39 (Deutsches Museum Archive Nl001/014, available online at http://www.Environmentandsociety.org/Exhibitions/Wegener-Diaries/Overview).

78. Ernst Sorge, "Winter at Eismitte," in *Greenland Journey, The Story of Wegener's German Expedition to Greenland in 1930 to 1931 as Told by Members of the Expedition and the Leader's Diary*, ed. Elsie Wegener and Fritz Loewe (London: Blackie & Sons, 1939), 180.
79. For analysis of Wegener's decision, and the bitter accusations and litigation leveled at Georgi in the aftermath of Wegener's death, see Murphy, *German Exploration of the Polar World: A History, 1870–1940*, 149ff. For a social analysis of the expedition, see Ursula Rack, *Sozialhistorische Studie zur Polarforschung Anhand von Deutschen und Oesterreich-Ungarischen Polarexpeditionen Zwischen 1868–1939* (Vienna: University of Vienna, 2010).
80. Wegener, "Tagebücher, April 1930-September 1930," p. 34 (Deutsches Museum Archive Nl001/014, available online at http://www.Environmentandsociety.org/Exhibitions/Wegener-Diaries/Overview).
81. Quoted in Wegener and Loewe, eds., *Greenland Journey, The Story of Wegener's German Expedition to Greenland in 1930 to 1931 as Told by Members of the Expedition and the Leader's Diary*, 113.
82. Wegener, "Tagebücher, April 1930-September 1930," p. 132.
83. Fritz Loewe, "The End of the Last Autumn Sledge Journey," in *Greenland Journey, The Story of Wegener's German Expedition to Greenland in 1930 to 1931 as Told by Members of the Expedition and the Leader's Diary*, ed. Elsie Wegener and Fritz Loewe (London: Blackie & Sons, 1939), 177.
84. Sorge, "Winter at Eismitte," 184.
85. Ernst Sorge, "Scientific Results of the Wegener Expedition to Greenland," *Geographical Journal* 81 (1933): 334.
86. Sorge, "Winter at Eismitte," 183.
87. Christian Kehrt, "Eternal Ice in the Cold War: The Polar Regions in Spatial and Environmental Perspective, 1957–1991" (Paper presented at the *Cold War Science, Colonial Politics and National Identity in the Arctic Workshop*, Aarhus, December 2010); Christian Kehrt, "Ponies, Dogs or Propeller Sledges? Alfred Wegener and the Limits of Modern Technology in Polar Exploration" (Paper presented at the SHOT Annual Meeting, Copenhagen, October 2012).
88. Kehrt, "Eternal Ice in the Cold War: The Polar Regions in Spatial and Environmental Perspective, 1957–1991," 3.
89. Sorge, "Scientific Results of the Wegener Expedition to Greenland," 335. Also see Alfred Wegener, *Mit Motorboot und Schlitten in Grönland* (Bielefeld and Leipzig: Verlag von Velhagen Klasing, 1930).
90. Quoted in Wegener and Loewe, eds., *Greenland Journey, The Story of Wegener's German Expedition to Greenland in 1930 to 1931 as Told by Members of the Expedition and the Leader's Diary*, 57–58.
91. Wegener, "Tagebücher, April 1930-September 1930," pp. 111–138.
92. Ibid., 138–147.
93. Georgi, *Mid-Ice: The Story of the Wegener Exhibition to Greenland*, 202.

NOTES

94. Johannes Georgi, "The First Sledge Journey Inland and the Establishment of the Eismitte Station," in *Greenland Journey, The Story of Wegener's German Expedition to Greenland in 1930 to 1931 as Told by Members of the Expedition and the Leader's Diary*, ed. Elsie Wegener and Fritz Loewe (London: Blackie & Sons, 1939), 95–96. This excavation can be seen in the silent film "Deutsche Grönlandexpedition Alfred Wegener 1931" (Reichsanstalt Für Film und Bild in Wissenschaft und Unterricht, Berlin, 1936, Bundesarchiv-Filmarchiv Berlin).
95. Georgi, "The First Sledge Journey Inland and the Establishment of the Eismitte Station," 93.
96. Sorge, "Winter at Eismitte," 179.
97. Quoted in Kehrt, "Eternal Ice in the Cold War: The Polar Regions in Spatial and Environmental Perspective, 1957–1991."
98. Sorge, "Winter at Eismitte," 192.
99. Georgi: Georgi, "The First Sledge Journey Inland and the Establishment of the Eismitte Station," 89. Sorge: quoted in Kehrt, "Eternal Ice in the Cold War: The Polar Regions in Spatial and Environmental Perspective, 1957–1991". Also, Georgi's diary and letters give an excellent portrait of daily life at Eismitte; they are preserved in Georgi, *Mid-Ice: The Story of the Wegener Exhibition to Greenland*.
100. Quoted in Voss, "Alfred Wegeners Weg als Polarforscher," 91.
101. Sorge, "Winter at Eismitte," 181, 94.
102. Ibid., 185–86.
103. Ibid., 188.
104. Quote: Michael Spender and Therkel Mathiassen, "Alfred Wegener's Greenland Expeditions 1929 and 1930–31: Review," *The Geographical Journal* 84 (1934): 518. The expedition's key scientific results appear in Kurt Wegener, *Wissenschaftliche Ergebnisse der Deutschen Grönland-Expedition Alfred Wegener 1929 und 1930/1931, Band I: Geschichte der Expedition* (Leipzig: F.A. Brockhaus, 1933).
105. Sorge, "Winter at Eismitte," 185.
106. François E. Matthes, "The Glacial Anticyclone Theory Examined in the Light of Recent Meteorological Data from Greenland, Part I," *American Geophysical Union Transactions* 27 (1946): 225. Georgi's meterological work at Eismitte also fed into the vibrant glacial anticyclone debate of the era; as well as Matthes see, for example, Johannes Georgi, "Greenland as a Switch for Cyclones," *Geographical Journal* 81 (1933); William H. Hobbs, "The Greenland Glacial Anticyclone," *Journal of Meteorology* 2 (1945).
107. Ahlmann, "Review of Scientific Results of the German Alfred Wegener Greenland Expedition 1929 and 1930–31," 135. For a similar evaluation, see "Review of Wissenschaftliche Ergebnisse der Deutschen Grönland-Expedition Alfred Wegener 1929 und 1930–31," *The Geographical Journal* 95, no. 5 (1940): 395.
108. See, for example, William H. Hobbs, "The Defense of Greenland," *Annals of the Association of American Geographers* 31, no. 2 (1941).

109. Matthes, "The Glacial Anticyclone Theory Examined in the Light of Recent Meteorological Data from Greenland, Part I," 324.
110. Henri Bader, "SIPRE Research Report 2, AD-014 366: Sorge's Law of Densification of Snow on High Polar Glaciers" (US Snow, Ice and Permafrost Research Establishment, 1953), 322.
111. Sorge, "Winter at Eismitte," 189.
112. "Scientific Results of the Wegener Expedition to Greenland," 339. Also see Ernst Sorge, "Glaziologische Untersuchungen in Eismitte," in *Wissenschaftliche Ergebnisse der Deutschen Groenland Expedition Alfred Wegener 1929 und 1930–31* (Leipzig: F.A. Brokaus, 1935).
113. Henri Bader, "United States Polar Ice and Snow Studies in the International Geophysical Year," in *Geophysics and the IGY*, ed. Hugh Odishaw and Stanley Ruttenberg (Washington, DC: American Geophysical Union, 1958), 178. Bader later gave this law a mathematical form and named it after Sorge. See Henri Bader, "Sorge's Law of Densification of Snow on High Polar Glaciers," *Journal of Glaciology* 2, no. 15 (1954).
114. Bernhard Brockamp, Ernst Sorge, and Kurt Wölcken, *Wissenschaftliche Ergebnisse der Deutschen Grönland-Expedition Alfred Wegener 1929 und 1930/1931, Band II: Seismik* (Leipzig: F.A. Brockhaus, 1933), 125ff; Sorge, "Scientific Results of the Wegener Expedition to Greenland," 337ff.
115. Spender and Mathiassen, "Alfred Wegener's Greenland Expeditions 1929 and 1930–31: Review," 520.
116. Sorge, "Scientific Results of the Wegener Expedition to Greenland," 335, 44.
117. Einar Storgaard, "Alfred Wegeners Grønlandsekspedition 1929–1931," *Geografisk Tidsskrift* 35, no. 4 (1932): 211.
118. Sorge, "Scientific Results of the Wegener Expedition to Greenland," 335.

2 Taming the Ice Sheet

1. Bernard Saladin d'Anglure, "Mauss et l'Anthropologie des Inuit," *Sociologie et Sociétés* 36, no. 2 (2004).
2. Paul-Emile Victor, *Mes Aventures Polaires* (Paris: Editions GP, 1975), 8. As a young anthropologist, Victor's two chief interests were Polynesia and the polar regions. While his working life centered on the polar world, he spent his final decades, from 1977 until his death in 1995, on Bora Bora in French Polynesia.
3. Paul-Emile Victor, *Boréal: La Joie dans la Nuit* (Paris: Grasset, 1938); Paul-Emile Victor, *Banquise: Le Jour Sans Ombre* (Paris: Grasset, 1939).
4. Geneviève De Lacour, "Retour au Groenland 70 Ans Après Paul-Emile Victor," *National Geographic French Edition*, March 2007, 4.
5. Robert Gessain, *Un Homme Marche Devant: La Dernière Traversée du Groenland en Traineaux à Chiens* (Paris: Arthaud, 1989); Eigil

Knuth, *Fire Mand og Solen: En Tur over Grønlands Indlandsis, 1936* (Copenhagen: Gyldendal, 1937).
6. The telegram appears in Gessain, *Un Homme Marche Devant: La Dernière Traversée du Groenland en Traineaux à Chiens*, 114.
7. Gérard Taylor, "Les Expéditions Polaires Francaises au Groënland, 1947" (Autres Expeditions Arctique 1947–1956 Folder, Expéditions Polaires Francaises 1914–2001 Collection (20110210-113), Archives Nationales de France, Fontainebleau), 2.
8. Victor, *Mes Aventures Polaires*, 56.
9. Paul-Emile Victor, "Je suis un Esquimau" (*Paris Soir*, 29 December 1937); Paul-Emile Victor, "Impressions du Bout du Monde," *Marianne*, 2 February 1938. Also see Paul-Emile Victor, "Les Jeux de Ficelle Chez les Eskimos d'Angmagssalik," *Journal de la Société des Américanistes* 29 (1937); Paul-Emile Victor, *My Eskimo Life* (London: Hamish Hamilton, 1939); Paul-Emile Victor, "Expéditions Polaires Francaises," *Atomes* 4 (1949).
10. Paul-Emile Victor, "The French Expedition to Greenland, 1948," *Arctic* 2 (1949): 135.
11. Taylor, "Les Expéditions Polaires Francaises au Groënland, 1947" (Autres Expéditions Arctique 1947–1956 Folder, Expéditions Polaires Francaises 1914–2001 Collection (20110210-113), Archives Nationales de France, Fontainebleau), 3–8. Also see Paul-Emile Victor, "Les Expéditions Francaises au Groënland," in *Revue Danoise*, Vol. 17, 1960 (LAC, RG85, Denmark and Greenland General File No. 1005–7, Vol. 14, Department of Northern Affairs and National Resources, ATIP Division Interim Box 46–2001099406).
12. While Victor initially only planned to extend his pre-war investigations in Greenland, he soon added the Antarctic to his agenda in order to gain political support and funding.
13. Paul-Emile Victor, "Projet Préliminaire des Explorations Polaires Francaises dans l'Arctique et dans l'Antarctique, 1947–1950, A—Synopsis, February 1947" (Organisation des EPF 1945–1964 Folder, Statuts EPF 1914–1959 Subfolder, Expéditions Polaires Francaises 1914–2001 Collection (20110210-001), Archives Nationales de France, Fontainebleau).
14. Jules Sébastian César Dumont d'Urville, *Voyage au Pôle Sud et dans l'Océanie sur les Corvettes l'Astrolabe et la Zélée, 1837–1840* (Paris: Gide et J. Baudry, 1854); Nicolas Skrotzky, *Terres Extrêmes: La Grande Aventure des Pôles* (Paris: Editions Denoël, 1986), 35ff.
15. Victor, "Projet Préliminaire des Explorations Polaires Francaises dans l'Arctique et dans l'Antarctique, 1947–1950, A—Synopsis, February 1947" (Organisation des EPF 1945–1964 Folder, Statuts EPF 1914–1959 Subfolder, Expéditions Polaires Francaises 1914–2001 Collection (20110210-001), Archives Nationales de France, Fontainebleau).
16. Australia and Britain were of particular concern as the French claim to Terre Adélie challenged Australia's Antarctic sector and antagonized

the British. See Peter J. Becks, *The International Politics of Antarctica* (London: Routledge, 1986), 29–30.
17. Paul-Emile Victor, "Letter to Ministère de l'Economie Nationale de la France, 19 May 1947" (Relations avec le Danemark 1947–1991 Folder, Expéditions Polaires Francaises 1914–2001 Collection (20110210–255), Archives Nationales de France, Fontainebleau).
18. Xavier Reppe, *Aurore sur l'Antarctique* (Paris: Nouvelles Editions Latines, 1957). As well as Terre Adélie, these interests also included the Kerguelen, Saint Paul, Amsterdam and Crozet Islands in the southern Indian Ocean and the Iles Eparses (Scattered Islands) near Madagascar.
19. "Expéditions Polaires Francaises, Missions Paul-Emile Victor: Terre Adélie, Groënland, 1947–1955: Rapport d'Activités," 5; "Rapport sur les Expéditions Polaires Francaises," 16 August 1983 (Folder 20110210–002, Expéditions Polaires Francaises 1914–2001 Collection, Archives Nationales de France, Fontainebleau), 1; Jean Malaurie, *Hummocks 1 et 2, Collection 'Terre Humaine'* (Paris: Plom, 1999[1955]), 32. The conversion to 2012 US dollars is made difficult by the rapid devaluation of the French franc after World War II.
20. "Note pour Mr P.E. Victor," 15 November 1947 (Organisation des EPF 1945–1964 Folder, Expéditions Polaires Francaises 1914–2001 Collection (20110210–001), Archives Nationales de France, Fontainebleau).
21. "Advertisements" (Autres Expéditions Arctique 1947–1956 Folder, Expéditions Polaires Francaises 1914–2001 Collection (20110210–113), Archives Nationales de France, Fontainebleau).
22. The leading work on the reconstruction of European science after World War II is John Krige, *American Hegemony and Postwar Reconstruction of Science in Europe* (Cambridge, MA: MIT Press, 2006). For France specifically, see Doris T. Zallen, "Louis Rapkine and the Restoration of French Science After the Second World War," *French Historical Studies* 17 (1991); Pierre Teissier, "Solid-State Chemistry in France: Structures and Dynamics of a Scientific Community Since World War II," *Historical Studies in the Natural Sciences* 40 (2010); Antoine Prost, *Les Origines de la Politique de la Recherche en France, 1939–1958* (Paris: Cahiers Pour l'Histoire du CNRS I, 1988). For France in the early postwar period more broadly, see William I. Hitchcock, *France Restored: Cold War Diplomacy and the Quest for Leadership in Europe, 1944–1954* (Chapel Hill, NC: University of North Carolina Press, 1998).
23. These plans were drawn up by the *Commissariat Général du Plan* (General Planning Commission), founded in 1946 by Charles de Gaulle, head of France's provisional postwar government.
24. Skrotzky, *Terres Extrêmes: La Grande Aventure des Pôles*, 72.
25. Lauge Koch, "Témoinages," 28 November 1951 (Organisation des EPF 1945–1964 Folder, Expéditions Polaires Francaises 1914–2001 Collection (20110210–001), Archives Nationales de France, Fontainebleau).

26. Paul-Emile Victor, *L'Iglou* (Paris: Stock, 1987).
27. For early French polar expeditions, see Louis Rey, *Groënland: Univers de Cristal* (Paris: Flammarion, 1974); Marthe Emmanuel, *La France et l'Exploration Polaire: De Verrazano à la Perouse, 1583–1788* (Paris: Nouvelles Editions Latines, 1959); Skrotzky, *Terres Extrêmes: La Grande Aventure des Pôles*.
28. "Expéditions Polaires Francaises, Missions Paul-Emile Victor: Terre Adélie, Groënland, 1947–1955: Rapport d'Activités," 53; Victor, "The French Expedition to Greenland, 1948," 136.
29. Michel Bouché, *Groënland: Station Centrale* (Paris: Bernard Grasset, 1952), 14–15.
30. "Twentieth Anniversary of Expéditions Polaires Francaises (Missions Paul-Emile Victor)," *Arctic* 21 (1968): 64.
31. "Expéditions Polaires Françaises, Missions Paul-Emile Victor, Expéditions Arctiques: Campagne Préparatoire au Groenland, 1948 (Rapports Préliminaires 5)" (Paris: Expéditions Polaires Françaises, 1954); "Liste des Observations et Travaux Scientifiques et Techniques des Expéditions au Groënland 1948 à 1953" (Autres Expéditions Arctique 1947–1956 Folder, Expéditions Polaires Francaises 1914–2001 Collection (20110210–113), Archives Nationales de France, Fontainebleau).
32. Robert Pommier, *Au-delà de Thule: Sur la Route des Glaces* (Paris: Amiot Dumont, 1953), 111.
33. "Expéditions Polaires Francaises, Missions Paul-Emile Victor: Terre Adélie, Groënland, 1947–1955: Rapport d'Activités"; Paul-Emile Victor et al., *Groënland: 1948–1949* (Paris: Arthaud, 1951), 27; Pierre Morel, "Un Déjeuner Avec Paul-Emile Victor et Roger Loubry, Projet d'Article pour Esso-Revue, Nouvelle Série," n.d. (Autres Expéditions Arctique 1947–1956 Folder, Expéditions Polaires Francaises 1914–2001 Collection (20110210–113), Archives Nationales de France, Fontainebleau), 10–11.
34. "Démarrage à Froid—Réchauffage," n.d. (Organisation des EPF 1945–1964 Folder, Equipements Subfolder, Expéditions Polaires Francaises 1914–2001 Collection (20110210–001), Archives Nationales de France, Fontainebleau).
35. Quoted in David Lampe, *Pyke, The Unknown Genius* (London: Evans Bros, 1959), 156.
36. D.J. Kinney, "Engineering Greenland: Icecap-1 and the Militarization of Arctic Technologies" (Paper presented at the SHOT Annual Meeting, Copenhagen, October 2012).
37. Victor et al., *Groënland: 1948–1949*, 18.
38. Ibid., 28.
39. Bouché, *Groënland: Station Centrale*, 55.
40. "Liste des Observations et Travaux Scientifiques et Techniques des Expéditions au Groënland 1948 à 1953" (Autres Expéditions Arctique 1947–1956 Folder, Expéditions Polaires Francaises 1914–2001

Collection (20110210-113), Archives Nationales de France, Fontainebleau), 22; "Expéditions Polaires Francaises, Missions Paul-Emile Victor: Terre Adélie, Groënland, 1947-1955: Rapport d'Activités"; Victor et al., *Groënland: 1948-1949.*
41. Bouché, *Groënland: Station Centrale*, 57.
42. Gaston Rouillon, "Ravitaillement par Air sur le Groënland, juillet-aout-octobre 1949" (Autres Expéditions Arctique 1947-1956 Folder, Expéditions Polaires Francaises 1914-2001 Collection (20110210-113), Archives Nationales de France, Fontainebleau), 3-4; "Expéditions Polaires Francaises, Missions Paul-Emile Victor: Terre Adélie, Groënland, 1947-1955: Rapport d'Activités," 63. The French were by no means the first to bring airplanes to the polar world, or to Greenland: in the interwar years, the Americans, Canadians, Danes and Soviets all used airplanes in the polar regions. The perception of precedence is part of a broader crafted narrative, discussed further at the end of this chapter.
43. See, for example, Georges de Caunes, "Comment Nous Avons Etabli un Pont 'Aérien' Entre l'Islande et le Groënland," *Paris-Presse*, 16 August 1949; Georges de Caunes, "Je Reviens du Grand Nord" (*Radiodiffusion Française*, 1949); Georges de Caunes, *Imarra: Aventures Groënlandaises* (Paris: Editions Hoebeke, 1998).
44. Rouillon, "Ravitaillement par Air sur le Groënland, juillet-aout-octobre 1949" (Autres Expéditions Arctique 1947-1956 Folder, Expéditions Polaires Francaises 1914-2001 Collection (20110210-113), Archives Nationales de France, Fontainebleau), 9.
45. Jean-Jacques Holtzscherer and Albert Bauer, "Expéditions Polaires Francaises, Missions Paul-Emile Victor, Expéditions Arctiques, Résultats Scientifiques: Contribution à la Connaissance de l'Inlandsis du Groënland, Mesures Seismiques et Synthèse Glaciologique (Communications Présentées à la Dixième Assemblée Générale de l'Union Géodesique et Géophysique Internationale Tenue à Rome en Septembre 1954)" (Paris: Expéditions Polaires Francaises, 1954), 11.
46. Victor et al., *Groënland: 1948-1949*, 20.
47. Bouché, *Groënland: Station Centrale*; Victor et al., *Groënland: 1948-1949*, 29.
48. Quote: Pommier, *Au-delà de Thule: Sur la Route des Glaces*, 119.
49. Mario Marret, "Réflexions Techniques," June 1953 (Organisation des EPF 1945-1964 Folder, Expéditions Polaires Francaises 1914-2001 Collection (20110210-001), Archives Nationales de France, Fontainebleau), 26-27; Raymond Latarjet, "Les Rations Alimentaires," *Atomes* 4 (1949).
50. Bouché, *Groënland: Station Centrale*, 84-99; "Expéditions Polaires Francaises, Missions Paul-Emile Victor: Terre Adélie, Groënland, 1947-1955: Rapport d'Activités," 136; "Liste des Observations et Travaux Scientifiques et Techniques des Expéditions au Groënland 1948 à 1953" (Autres Expéditions Arctique 1947-1956 Folder,

Expéditions Polaires Francaises 1914–2001 Collection (20110210-113), Archives Nationales de France, Fontainebleau), 23.
51. Bouché, *Groënland: Station Centrale*, 98.
52. Ibid., 61, 97–100. This emphasis on physical and mental comforts continued to be important in later EPF expeditions. Writing in 1953, Marret noted that expedition teams were too busy to wash their clothing by hand and argued that automatic washing machines would increase the efficiency of expeditions: "at the risk of making you smile," he wrote, "let it be known that we consider a washing machine to be indispensable." Marret also commented on the aesthetics of polar stations: "why neglect interior decoration, which can be done with so few materials and which has such a strong psychological effect? What are a few kilos of wallpaper and glue, or of paint, against the pleasure which a pleasant interior brings? How difficult is it for the dishes to be the same as those the team members might eat off at their homes? For the drinking glasses to be real glasses and not metallic cans? We cannot attach too much importance to these details which neither cost money nor time, but only a little bit of ingenuity, and which bring so much well-being." Marret, "Réflexions Techniques," June 1953 (Organisation des EPF 1945–1964 Folder, Expéditions Polaires Francaises 1914–2001 Collection (20110210-001), Archives Nationales de France, Fontainebleau), 20–30.
53. Quoted in Christian Kehrt, "Eternal Ice in the Cold War: The Polar Regions in Spatial and Environmental Perspective, 1957–1991" (Paper presented at the *Cold War Science, Colonial Politics and National Identity in the Arctic Workshop*, Aarhus, December 2010), 5. Also see Victor, "The French Expedition to Greenland, 1948"; "Expéditions Polaires Francaises, Missions Paul-Emile Victor: Terre Adélie, Groënland, 1947–1955: Rapport d'Activités."
54. For the historical development of weather balloons and upper atmosphere probing, see Frederik Nebeker, *Calculating the Weather: Meteorology in the 20th Century* (San Diego: Academic Press, 1995).
55. Bouché, *Groënland: Station Centrale*, iii, 87.
56. Janet Martin-Nielsen, "'The Deepest and Most Rewarding Hole Ever Drilled': Ice Cores and the Cold War in Greenland," *Annals of Science* 70 (2013); Maiken Lolck, *Klima, Kold Krig og Iskerner* (Aarhus: Aarhus Universitetforlag, 2006); Aant Elzinga, "Some Aspects in the History of Ice Core Drilling and Science from IGY to EPICA" (3rd Scar Antarctic History Action Group Workshop, Byrd Polar Research Center, Ohio State University, October 2007).
57. André Cailleux and Emile Thellier, "Rapport Provisoire sur un Complement à l'Etude de l'Inlandsis Groënlandais ou Antarctique: Etude de la Stratification de la Neige et des Particules Incluses," 15 May 1947 (Organisation des EPF 1945–1964 Folder, Expéditions Polaires Francaises 1914–2001 Collection (20110210-001), Archives Nationales de France, Fontainebleau).

58. André Cailleux, "Premiers Enseignements Glaciologiques des Expéditions Polaires Francaises, 1948–1951" (Expéditions Polaires Francaises, 1951); Jean-Charles Heuberger, *Expéditions Polaires Francaises, Missions Paul-Emile Victor, V: Glaciologie Groënland, Volume 1, Forages sur l'Inlandsis* (Paris: Hermann et Cie, 1954); Holtzscherer and Bauer, "Expéditions Polaires Francaises, Missions Paul-Emile Victor, Expéditions Arctiques, Résultats Scientifiques: Contribution à la Connaissance de l'Inlandsis du Groënland, Mesures Seismiques et Synthèse Glaciologique."
59. Hans W. Ahlmann, "The Contribution of Polar Expeditions to the Science of Glaciology (A Lecture Delivered to the Scott Polar Research Institute on 1 May 1948)," *Polar Record* 5 (1949).
60. Holtzscherer and Bauer, "Expéditions Polaires Francaises, Missions Paul-Emile Victor, Expéditions Arctiques, Résultats Scientifiques: Contribution à la Connaissance de l'Inlandsis du Groënland, Mesures Seismiques et Synthèse Glaciologique"; "Expéditions Polaires Francaises, Missions Paul-Emile Victor: Terre Adélie, Groënland, 1947–1955: Rapport d'Activités." The fate of these markers is discussed in chapter 4.
61. Georges Dubois, "Données Numériques Relatives aux Glaciations Quaternaires," *Bulletin de l'Association de l'Institut des Sciences Géologiques de Strasbourg* (1931); Richard F. Flint, *Glacial Geology and the Pleistocene Epoch* (New York: John Wiley, 1948).
62. Holtzscherer and Bauer, "Expéditions Polaires Francaises, Missions Paul-Emile Victor, Expéditions Arctiques, Résultats Scientifiques: Contribution à la Connaissance de l'Inlandsis du Groënland, Mesures Seismiques et Synthèse Glaciologique," 24.
63. Edward Muller, "The Top of the World," *The Reader's Digest*, March 1954, 159; Viggo Kampmann, "Le Danemark Arctique," in *Revue Danoise*, Vol. 17, 1960 (LAC, RG85, Denmark and Greenland General File No. 1005-7, Vol. 14, Department of Northern Affairs and National Resources, ATIP Division Interim Box 46–2001099406), U1.
64. Hans W. Ahlmann, "Letter to Paul-Emile Victor, 12 June 1954" (Organisation des EPF 1945–1964 Folder, Expéditions Polaires Francaises 1914–2001 Collection (20110210–001), Archives Nationales de France, Fontainebleau).
65. Lauge Koch, "Letter to Paul-Emile Victor, 28 November 1951" (Organisation des EPF 1945–1964 Folder, Expéditions Polaires Francaises 1914–2001 Collection (20110210–001), Archives Nationales de France, Fontainebleau). Also see J. Kanwisher, "Letter to Albert Bauer, 17 May 1956" (Organisation EGIG: Projets, Réunions Préliminaires 1954–1957 Folder, Expéditions Polaires Francaises 1914–2001 Collection (20110210–117), Archives Nationales de France, Fontainebleau); Bernhard Brockamp, "Letter to Victor, 21 July 1952" (Autres Expéditions Arctique 1947–1956 Folder, Expéditions

NOTES 143

Polaires Francaises 1914–2001 Collection (20110210–113), Archives Nationales de France, Fontainebleau).
66. The political situation of Greenland during and after World War II is discussed in the next chapter.
67. See Peder Roberts, "Nordic or National? Postwar Visions of Polar Conflict and Cooperation," in *Science, Geopolitics and Culture in the Polar Regions: Norden Beyond Borders*, ed. Sverker Sörlin (Surrey, England: Ashgate, 2013). Danish scientific nationalism is discussed further in the epilogue.
68. Hans W. Ahlmann, "Letter to Paul-Emile Victor, 29 July 1947" (Organisation des EPF 1945–1964 Folder, Expéditions Polaires Francaises 1914–2001 Collection (20110210–001), Archives Nationales de France, Fontainebleau).
69. Eigil Knuth, "Letter to Paul-Emile Victor, 15 May 1948" (Relations avec le Danemark 1947–1991 Folder, Expéditions Polaires Francaises 1914–2001 Collection (20110210–255), Archives Nationales de France, Fontainebleau).
70. Dansk Udenrigsministeriet, "Note Verbale to Ambassade de France au Danemark, 7 May 1947, Copenhagen" (Relations avec le Danemark 1947–1991 Folder, Expéditions Polaires Francaises 1914–2001 Collection (20110210–255), Archives Nationales de France, Fontainebleau).
71. Paul-Emile Victor, "Letter to Eske Brun, 19 May 1947" (Relations avec le Danemark 1947–1991 Folder, Expéditions Polaires Francaises 1914–2001 Collection (20110210–255), Archives Nationales de France, Fontainebleau); Paul-Emile Victor, "Letter to Eske Brun, 28 May 1947" (Relations avec le Danemark 1947–1991 Folder, Expéditions Polaires Francaises 1914–2001 Collection (20110210–255), Archives Nationales de France, Fontainebleau); Paul-Emile Victor, "Letter to Eske Brun, 13 January 1948" (Relations avec le Danemark 1947–1991 Folder, Expéditions Polaires Francaises 1914–2001 Collection (20110210–255), Archives Nationales de France, Fontainebleau).
72. Eske Brun, "Letter to Paul-Emile Victor, 22 October 1947" (Relations avec le Danemark 1947–1991 Folder, Expéditions Polaires Francaises 1914–2001 Collection (20110210–255), Archives Nationales de France, Fontainebleau). The Commission for Scientific Research in Greenland (in Danish, *Kommissionen for Videnskabelige Undersøgelser i Grønland*, and in Greenlandic, *Kalaallit Nunaanni Ilisimatusarneq Pillugu Isumalioqatigiissitamut*) was founded in 1878 to coordinate and advise on scientific activities in Greenland. The commission's press, *Meddelelser om Grønland*, is a valuable source of information about scientific research on the island.
73. Ambassade de la République Francaise au Danemark, "Letter to Paul-Emile Victor, 2 March 1948, Communication No. 158" (Relations avec le Danemark 1947–1991 Folder, Expéditions Polaires Francaises

1914–2001 Collection (20110210–255), Archives Nationales de France, Fontainebleau). The French contributed 40,000 kroner to an expense account at the Greenland Office in 1948. They continued to contribute smaller sums in the following years, and the money was spent by the Danes on items such as postage, telegrams, and the transportation of goods and people. See Grønlandsdepartementet, "Statement of Account, Expéditions Polaires Francaises" (Relations avec le Danemark 1947–1991 Folder, Expéditions Polaires Francaises 1914–2001 Collection (20110210–255), Archives Nationales de France, Fontainebleau).

74. Grønlands Styrelse, "Letter to Paul-Emile Victor, 25 February 1948" (Relations avec le Danemark 1947–1991 Folder, Expéditions Polaires Francaises 1914–2001 Collection (20110210–255), Archives Nationales de France, Fontainebleau).

75. Paul-Emile Victor, "Letter to Eske Brun, 3 March 1949" (Relations avec le Danemark 1947–1991 Folder, Expéditions Polaires Francaises 1914–2001 Collection (20110210–255), Archives Nationales de France, Fontainebleau); Eske Brun, "Letter to Paul-Emile Victor, 22 December 1948" (Relations avec le Danemark 1947–1991 Folder, Expéditions Polaires Francaises 1914–2001 Collection (20110210–255), Archives Nationales de France, Fontainebleau).

76. Eske Brun, "Letter to Paul-Emile Victor, 18 February 1949, Jrn. Nr. 11664/47" (Relations avec le Danemark 1947–1991 Folder, Expéditions Polaires Francaises 1914–2001 Collection (20110210–255), Archives Nationales de France, Fontainebleau); Paul-Emile Victor, "Letter to Eske Brun, 18 March 1949" (Relations avec le Danemark 1947–1991 Folder, Expéditions Polaires Francaises 1914–2001 Collection (20110210–255), Archives Nationales de France, Fontainebleau).

77. Gaston Rouillon, "Compte-Rendu sur l'Accident du Mont Forel," 4 August 1951 (Relations avec le Danemark 1947–1991 Folder, Expéditions Polaires Francaises 1914–2001 Collection (20110210–255), Archives Nationales de France, Fontainebleau); Paul-Emile Victor and Jean Vaugelade, "Letter to Eske Brun, 9 August 1951" (Relations avec le Danemark 1947–1991 Folder, Expéditions Polaires Francaises 1914–2001 Collection (20110210–255), Archives Nationales de France, Fontainebleau).

78. Paul-Emile Victor, "Letter to the Family of Jens Jarl, 9 August 1951" (Relations avec le Danemark 1947–1991 Folder, Expéditions Polaires Francaises 1914–2001 Collection (20110210–255), Archives Nationales de France, Fontainebleau).

79. Rouillon, "Compte-Rendu sur l'Accident du Mont Forel," 4 August 1951 (Relations avec le Danemark 1947–1991 Folder, Expéditions Polaires Francaises 1914–2001 Collection (20110210–255), Archives Nationales de France, Fontainebleau), 37–38.

80. Victor, "Letter to Ministère de l'Economie Nationale de la France, 19 May 1947" (Relations avec le Danemark 1947–1991 Folder, Expéditions Polaires Francaises 1914–2001 Collection (20110210–255), Archives Nationales de France, Fontainebleau).
81. "La France dans les Régions Polaires," November 1955 (Folder 20110210–002, Expéditions Polaires Francaises 1914–2001 Collection, Archives Nationales de France, Fontainebleau), 2.
82. "Twentieth Anniversary of Expéditions Polaires Francaises (Missions Paul-Emile Victor)," 60. Also see "Rapport Général d'Activités Arctiques et Antarctiques Depuis 1947," December 1970 (Expéditions Franco-Americaines au Groënland 1952–1956 Folder, Expéditions Polaires Francaises 1914–2001 Collection (20110210–015), Archives Nationales de France, Fontainebleau), 3; "Les Expéditions Francaises au Groënland," in *Revue Danoise*, Vol. 17, 6–10 (LAC, RG85, Denmark and Greenland General File No. 1005–7, Vol. 14, Department of Northern Affairs and National Resources, ATIP Division Interim Box 46–2001099406), 6.
83. Marret, "Réflexions Techniques," June 1953 (Organisation des EPF 1945–1964 Folder, Expéditions Polaires Francaises 1914–2001 Collection (20110210–001), Archives Nationales de France, Fontainebleau), 2.
84. "La Guerre à 40° au Dessous de Zéro," n.d. (Organisation des EPF Folder, Expéditions Polaires Francaises 1914–2001 Collection (20110210–003), Archives Nationales de France, Fontainebleau); "Groenland—Le Danemark Arctique, Conseiller Culturel, Ambassade de Danemark à Paris," 27 October 1960 (Relations avec le Danemark 1947–1991 Folder, Expéditions Polaires Francaises 1914–2001 Collection (20110210–255), Archives Nationales de France, Fontainebleau); "Exposition: Le Froid et ses Merveilles, 1962" (Rigsarkivet, 0030, Grønlandsministeriet Collection, Journalsager 1957–1989, Box 2083, Folder Paul-Emile Victor, Ekspeditioner, 1420-21-01).
85. Ambassade de France (New York), "Letter to Jacques Cortadellas, 8 May 1950, No. 7744" (Relations Internationales: Etats-Unis, 1947–1983 Folder, Expéditions Polaires Francaises 1914–2001 Collection (20110210–256), Archives Nationales de France, Fontainebleau).
86. See Paul-Emile Victor, *Apoutsiak, Le Petit Flocon de Neige* (Paris: Flammarion, 1948).
87. "Images d'un Eté" (Un Film des Expéditions Polaires Francaises, Missions Paul-Emile Victor, produced by Armor-Films, 1950); Mario Marret and Fred Orani, "Nous Avons Vingt Ans" (Film produced by La Société Nouvelle Armor-Films, 1968).
88. Marcel Ichac and Jean-Jacques Languepin, "Groënland, 20,000 Lieues sur les Glaces" (Film: Shot 1949, Released 1952).
89. Rey, *Groënland: Univers de Cristal*, 257–58.

90. Børge Fristrup, *The Greenland Ice Cap*, trans. David Stoner (Copenhagen: Rhodos, 1966), 130–31.
91. de Caunes, "Comment Nous Avons Etabli un Pont 'Aérien' Entre l'Islande et le Groënland."
92. Leonard A. LeSchack, "The French Polar Effort and the Expéditions Polaires Francaises," *Arctic* 17 (1964): 13.
93. "Twentieth Anniversary of Expéditions Polaires Francaises (Missions Paul-Emile Victor)," 61–62. The US military used EPF as technical and scientific consultants both stateside and in Greenland between 1952 and 1958 (cf. chapter 3). In 1961, the USSR Academy of Science asked EPF to participate in a glaciological study of the Antarctic platform between Mirny and Vostok stations. This cooperation later resulted in the joint EPF-Expéditions Antarctiques Soviétiques expeditions of 1963–1964 and 1968–1969.
94. "Expéditions Polaires Francaises, Missions Paul-Emile Victor: Terre Adélie, Groënland, 1947–1955: Rapport d'Activités," 5. For EPF's internal problems (including personal disagreements and financial troubles), see Thierry Fournier, "Chapitre VI: Paul-Emile Victor et les Expéditions Américaines au Groënland, 1952–1957" (Unpublished PhD thesis chapter, Ecole Nationale des Chartes, France, 2012).
95. The base, named Base Dumont d'Urville, was built in 1956 and has been permanently occupied since 1959. In this guise, EPF worked closely with the new territorial office, *Terres Australes et Antarctiques Françaises* (French Southern and Antarctic Lands), which replaced the National Center for Scientific Research as the main funding body for EPF's work in the southern hemisphere. The classic source for international politics and the Antarctic is Becks, *The International Politics of Antarctica*.

3 The Longest Trek

1. This opening section is based on the author's two recent papers: Janet Martin-Nielsen, "The Other Cold War: The United States and Greenland's Ice Sheet Environment, 1948–1966," *Journal of Historical Geography* 38 (2012); and Janet Martin-Nielsen, "'The Deepest and Most Rewarding Hole Ever Drilled': Ice Cores and the Cold War in Greenland," *Annals of Science* 70 (2013).
2. Emil G. Beaudry, "Air Potentialities of the Greenland Ice Cap in High Latitude Defense: A Research Paper Submitted to the Faculty of the Air Command and Staff School of the Air University, Code No. 12, Maxwell Air Force Base, Alabama," Originally Secret, 1949 (NARA, RG 319, Records of the Army Staff), 1–24. Similar sentiments are also expressed in higher-level documents such as "Military Requirements for Base Rights, Joint Strategic Plans Committee 684/52, 23 March 1949," Originally Top Secret (NARA, RG 218, Section 36, CCS 360).

3. See Sanjay Chaturvedi, *The Polar Regions: A Political Geography* (Chichester, UK: John Wiley & Sons, 1996); Fae L. Korsmo, "The Early Cold War and US Arctic Research," in *Extremes: Oceanography's Adventures at the Poles*, ed. Keith R. Benson and Helen M. Rozwadowski (Sagamore Beach, MA: Science History Publications, 2007); Fae L. Korsmo, "Glaciology, the Arctic, and the US Military, 1945–58," in *New Spaces of Exploration: Geographies of Discovery in the 20th Century*, ed. Simon Naylor and James R. Ryan (London: I.B. Tauris, 2010).
4. "Appendix 1: Importance of the High Arctic to North American Defense," in *Report of the Arctic Institute of North America, Presented at the Hearings Before the Committee on Merchant Marine and Fisheries, House of Representatives 85th Congress, 2nd Session, 22–24 January 1958* (New York: Office of Naval Research, Arctic Research Advisory Committee, 25 October 1957), 40–41. Also see "US Objectives with Respect to the USSR to Counter Soviet Threats to US Security, NSC 20/4, 23 November 1948," Originally Top Secret, in *Foreign Relations of the United States* (Washington, DC: Government Printing Office, Department of State); "Byrd Stresses Use of Arctic in a War," *The New York Times* (November 18, 1947). For the other side of the coin—that is, Soviet Arctic strategy during the Cold War—see Terence Armstrong, *The Russians in the Arctic: Aspects of Soviet Exploration and Exploitation of the Far North, 1937–57* (Westport, CT: Greenwood Press, 1958); Ellman Ellingsen, "The Military Balance on the Northern Flank," in *Clash in the North: Polar Summitry and NATO's Northern Flank*, ed. Walter Goldstein (Washington, DC: Pergamon-Brassey's, 1988); Olli-Pekka Jalonen, "The Strategic Significance of the Arctic," in *The Arctic Challenge: Nordic and Canadian Approaches to Security and Cooperation in an Emerging International Region*, ed. Kari Möttölä (Boulder, CO: Westview Press, 1988).
5. "PJV Journal," 27 February 1960 (Rigsarkivet, Udenrigsministeriet Collection, 105.F.10).
6. "North Atlantic Military Committee Standing Group: Final Decision on SG 161/15—A Report by the Standing Group on the Soviet Bloc Strength and Capabilities, 11 May 1962," Originally Cosmic Top Secret (NATO, SG 161/15). Also see "Report by the Standing Group to the Military Committee, NATO Medium Term Defense Plan, 1 July 1954," Originally Secret (NATO, SG 20/2, Parts I & III); "Note by the Secretary to the Standing Group on Report by the Ad Hoc Committee on Atlantic Ocean Air Traffic Control, 11 March 1954," Originally Cosmic Top Secret (NATO, SG 230/1); "Long-Term Scientific Studies for the Standing Group North Atlantic Treaty Organization by Working Group III on Geophysics (Von Karman Committee), March 1961," Originally Secret (NATO, VKC-EX1-GPIII).

7. "Record of Meeting of the North Atlantic Military Committee, Held at 0930, Monday, 22 November 1954, in the International Conference Room, State Dept. Bldg. SA-17, Washington," Originally Secret (NATO, MC 10th Session, 29 November 1954, 1–48).
8. Beaudry, "Air Potentialities of the Greenland Ice Cap in High Latitude Defense: A Research Paper Submitted to the Faculty of the Air Command and Staff School of the Air University, Code No. 12, Maxwell Air Force Base, Alabama," Originally Secret (NARA, RG 319, Records of the Army Staff), 9. After this rescue, Beaudry was awarded the US Air Force's McKay trophy for the most meritorious military flight of 1948 and celebrated in the popular press in articles such as "Honored for Most Meritorious Flight", *The New York Times* (December 21, 1949) and "Heroes: Welcome Home", *Time Magazine* (January 10, 1949).
9. Bruce C. Paton, "Cold, Casualties, and Conquests: The Effects of Cold on Warfare," in *Medical Aspects of Harsh Environments 1*, ed. Kent B. Pandolf and Robert E. Burr (Falls Church, VA: Office of The Surgeon General, US Army, 2001), 335.
10. Hanson W. Baldwin, "Arctic Repels Warfare: US Exercises Show Great Difficulties, Make Large-Scale Operations Unlikely," *The New York Times* (February 8, 1948). Cold-weather injuries and disasters in this context are elaborated on in Ken Coates and William R. Morrison, *The Alaska Highway in World War II: The US Army of Occupation in Canada's Northwest* (Toronto: University of Toronto Press, 1992); Matthew Farish, "Creating Cold War Climates: The Laboratories of American Globalism," in *Environmental History and the Cold War*, ed. John R. McNeill and Corinna R. Unger (Cambridge: Cambridge University Press, 2010); Korsmo, "The Early Cold War and US Arctic Research"; Paton, "Cold, Casualties, and Conquests: The Effects of Cold on Warfare"; Norman F. Washburne, "A Survey of Human Factors in Military Performance in Extreme Cold Weather, Report AD 477889" (The George Washington University Human Resources Research Office/Department of the Army, 1960); Tom Whayne, *Cold Injury in World War II: A Study in the Epidemiology of Trauma* (PhD Thesis, Harvard School of Public Health, 1950).
11. Baldwin, "Arctic Repels Warfare: US Exercises Show Great Difficulties, Make Large-Scale Operations Unlikely".
12. The leading works on the rise of the geophysical sciences in the US in the postwar era are Ronald E. Doel, "Constituting the Postwar Earth Sciences: The Military's Influence on the Environmental Sciences in the USA after 1945," *Social Studies of Science* 33 (2003); Ronald E. Doel, "Quelle Place Pour les Sciences de l'Environnement Physique dans l'Histoire Environnementale?," *Revue d'Histoire Moderne et Contemporaine* 56 (2009). For glaciology, also see Korsmo, "The Early Cold War and US Arctic Research"; Korsmo, "Glaciology, the Arctic, and the US Military, 1945–58".

13. "Appendix 1: Importance of the High Arctic to North American Defense," in *Report of the Arctic Institute of North America, Presented at the Hearings Before the Committee on Merchant Marine and Fisheries, House of Representatives 85th Congress, 2nd Session, 22–24 January 1958*, 41.
14. "Appendix 1: Importance of the High Arctic to North American Defense," in *Report of the Arctic Institute of North America, Presented at the Hearings Before the Committee on Merchant Marine and Fisheries, House of Representatives 85th Congress, 2nd Session, 22–24 January 1958*, 38. Emphasis added. Science and geography were also important in other cold environments, including Alaska, northern Canada and Antarctica, during the Cold War. The cases of Alaska and northern Canada are addressed in Matthew Farish, "The Lab and the Land: Overcoming the Arctic in Cold War Alaska," *Isis* 104 (2013); Laurel J. Hummel, "The US Military as Geographical Agent: The Case of Cold War Alaska," *The Geographical Review* 95 (2005); P. Whitney Lackenbauer and Matthew Farish, "The Cold War on Canadian Soil: Militarizing a Northern Environment," *Environmental History* 12 (2007); Richard C. Powell, "Science, Sovereignty and Nation: Canada and the Legacy of the International Geophysical Year, 1957–1958," *Journal of Historical Geography* 34 (2008). The case of Antarctica is addressed in Dian O. Belanger, *Deep Freeze: The United States, the International Geophysical Year, and the Origins of Antarctica's Age of Science* (Boulder, CO: University of Colorado Press, 2006); Christy Collis and Klaus Dodds, "Assault on the Unknown: The Historical and Political Geographies of the International Geophysical Year, 1957–1958," *Journal of Historical Geography* 34 (2008); Simon Naylor, Katrina Dean, and Martin Siegert, "The IGY and the Ice Sheet: Surveying Antarctica," *Journal of Historical Geography* 34 (2008); Simone Turchetti et al., "On Thick Ice: Scientific Internationalism and Antarctic Affairs, 1957–1980," *History and Technology* 24 (2008).
15. For overviews of US Cold War scientific research in Greenland, see Heymann et al., "Exploring Greenland: Science and Technology in Cold War Settings," *Scientia Canadensis* 33 (2010): 11–42; Owen Wilkes and Jan Øberg, *Military Research and Development in Denmark and Greenland* (Lund, Sweden: Lund University, 1982). For studies of the geophysical sciences in Cold War Greenland, see Henrik Knudsen, "Cold War, Ionospheric Research in Greenland, and the Politics of Rockets: A Study of the Ill-Fated Operation PCA 68" (forthcoming); Christopher Jacob Ries, "On Frozen Ground: William E. Davies and the Military Geology of Northern Greenland, 1952–1960," *The Polar Journal* 2, no. 2 (2012); Martin-Nielsen, "The Other Cold War: The United States and Greenland's Ice Sheet Environment, 1948–1966"; Martin-Nielsen, "'The Deepest and Most Rewarding Hole Ever Drilled': Ice Cores and the Cold War in Greenland".

16. H.F.C. Hannis, "General Orders No. 2: Establishment of the Snow, Ice and Permafrost Research Establishment, Department of the Army, Washington DC, 9 March 1949" (Arktisk Institut, Box SIPRE Reports 1–8), 19.
17. "SIPRE Report 1, 1950" (Arktisk Institut, Box SIPRE Reports 1–8), 35.
18. Carl S. Benson and Charles R. Wilson, "Barry Bishop's Research on the Sheer Moraines in the Thule Area, Northwest Greenland," *Mountain Research and Development* 16, no. 3 (1996): 310.
19. Richard F. Flint, "Snow, Ice and Permafrost in Military Operations, 1950" (Arktisk Institut, Box SIPRE Reports 1–8), 3–5.
20. Børge Fristrup, *The Greenland Ice Cap*, trans. David Stoner (Copenhagen: Rhodos, 1966), 146.
21. Korsmo, "Glaciology, the Arctic, and the US Military, 1945–58", 126, 46.
22. Richard P. Goldthwait, "Glaciology: Report of the University Committee on Polar Research, National Academy of Sciences, 1956" (LAC, Arctic Institute of North America Fonds, MG28, I79, Volume 212, Folder 29), 22–27. Goldthwait went on to found Ohio State University's Institute of Polar Studies (since re-named the Byrd Polar Research Center in honor of US Rear Admiral and polar explorer Richard E. Byrd), one of the leading glaciological institutes in the US.
23. Quoted in Korsmo, "The Early Cold War and US Arctic Research", 180.
24. Known as the 'city under the ice', Camp Century was a 225-person nuclear-powered American 'city' buried inside the Greenland ice sheet. It stands out as one of the most incredible US military facilities of the Cold War. See Charles Michael Daugherty, *City Under the Ice: The Story of Camp Century* (New York: The Macmillan Company, 1963); Walter Wager, *Camp Century: City Under the Ice* (Philadelphia, PA: Chilton Books, 1962).
25. For the Snowman Project, see "The Snowman Project, HQ Atlanta Division, ATC, Westover Field, Massachusetts, 1947" (NARA, RG 341, Air Force: Plans, Project Decimal File 1942–1954, Box 831, SG 581, TS). For Project Mint Julep, see "ADTIC Publication A-104a: Project Mint Julep—Investigation of Smooth Areas of the Greenland Ice Cap, 1953, Part 1" (Research Studies Institute, Air University, Maxwell Air Force Base, Alabama: Arctic, Desert, Tropic Information Center, 1955). For Operation Icecap, see Robert L. Nichols, "Scientific Studies on the Ice Cap and in Inglefield Land with Special Reference to Military Significance, Final Report on the Scientific Program B, Operation Ice Cap 1953" (Project DA 9–98–07–002, Stanford Research Institute, 1954). For Operation King Dog, see Chester C. Langway, "CRREL Report SR 31: Snow Studies and Other Observations: Operation King Dog, Sondrestrom, Greenland" (US Cold Regions Research and Engineering Laboratory, 1959).
26. Carl S. Benson, "Oral History Interview with Karen Brewster" (Byrd Polar Research Center Oral History Program, June 22, 2001), 10.

NOTES

27. Carl S. Benson, "Stratigraphic Studies in the Snow and Firn of the Greenland Ice Sheet" (California Institute of Technology, 1960), ii.
28. Robert W. Christie, "Pioneer in Pathology," *Dartmouth Medicine* 27 (2003).
29. "Organic Act of the US Geological Survey, US Statutes at Large, V. 20" (1879), 394. For the Geological Survey, also see Mary C. Rabbitt, "The United States Geological Survey, 1879–1989," US Geological Survey Circular 1050 (US Government Printing Office, 1989); Clifford N. Nelson, "Geological Survey," in *The History of Science in the United States: An Encyclopedia*, ed. Marc Rothenberg (New York & London: Garland, 2001); A. Hunter Dupree, *Science in the Federal Government: A History of Policies and Activities to 1940* (Cambridge, MA: Harvard University Press, 1957).
30. Benson, "Oral History Interview with Karen Brewster," 5, 49.
31. Robert Pommier, *Au-delà de Thule: Sur la Route des Glaces* (Paris: Amiot Dumont, 1953), 50.
32. Benson and Wilson, "Barry Bishop's Research on the Sheer Moraines in the Thule Area, Northwest Greenland," 310–11.
33. Fristrup, *The Greenland Ice Cap*, 155.
34. Benson, "Stratigraphic Studies in the Snow and Firn of the Greenland Ice Sheet," iii.
35. Benson, "Oral History Interview with Karen Brewster," 60.
36. Carl S. Benson, "Letter to Paul-Emile Victor, 20 October 1955" (Relations Internationales: Etats-Unis, 1947–1983 Folder, Expéditions Polaires Francaises 1914–2001 Collection (20110210–256), Archives Nationales de France, Fontainebleau).
37. Benson, "Stratigraphic Studies in the Snow and Firn of the Greenland Ice Sheet," iii.
38. Benson, "Oral History Interview with Karen Brewster," 8.
39. Gaston Rouillon and Paul-Emile Victor, "Letter to Henri Bader, 7 January 1955" (Relations Internationales: Etats-Unis, 1947–1983 Folder, Expéditions Polaires Francaises 1914–2001 Collection (20110210–256), Archives Nationales de France, Fontainebleau).
40. Benson, "Stratigraphic Studies in the Snow and Firn of the Greenland Ice Sheet," 9.
41. Carl S. Benson and Richard H. Ragle, "SIPRE Special Report 18: Project Jello, SIPRE Greenland Expedition 1955, Report on Special Foods Provided by the Quartermaster Food and Container Institute" (US Snow, Ice and Permafrost Research Establishment, 1956), 3.
42. Ibid., 3–8.
43. Ibid., I-8.
44. Benson, "Oral History Interview with Karen Brewster," 30.
45. Benson and Ragle, "SIPRE Special Report 18: Project Jello, SIPRE Greenland Expedition 1955, Report on Special Foods Provided by the Quartermaster Food and Container Institute," 7. For Party Crystal, see Carl S. Benson, "SIPRE Report 25: Operations and Logistics of

Ice-Cap Party Crystal (April 1955)," in *Corps of Engineers, Greenland Ice Cap Research Program, Studies Completed in 1954* (Vicksburg, Miss.: Army-MRC, 1955); Carl S. Benson, "SIPRE Report 24: Scientific Work of Party Crystal, 1954" (Preliminary Report, April 1955), in *Corps of Engineers, Greenland Ice Cap Research Program, Studies Completed in 1954* (Vicksburg, MS: Army-MRC, 1955).
46. Franz A. Koehler, *Special Rations for the Armed Forces, 1946–53* (Washington DC: Office of the Quartermaster General, 1958).
47. First quote: Benson and Ragle, "SIPRE Special Report 18: Project Jello, SIPRE Greenland Expedition 1955, Report on Special Foods Provided by the Quartermaster Food and Container Institute," 7. Second quote: Benson, "Oral History Interview with Karen Brewster," 30.
48. Benson, "Stratigraphic Studies in the Snow and Firn of the Greenland Ice Sheet," iii.
49. Benson and Ragle, "SIPRE Special Report 18: Project Jello, SIPRE Greenland Expedition 1955, Report on Special Foods Provided by the Quartermaster Food and Container Institute," 9.
50. Ibid., I-2.
51. Ragle, Christie and Skinrood's comments appear in Ibid., I-3—I-30.
52. "ADTIC Publication A-104a: Project Mint Julep—Investigation of Smooth Areas of the Greenland Ice Cap, 1953, Part 1," v–vi.
53. Benson and Ragle, "SIPRE Special Report 18: Project Jello, SIPRE Greenland Expedition 1955, Report on Special Foods Provided by the Quartermaster Food and Container Institute," 7.
54. "ADTIC Publication A-104a: Project Mint Julep—Investigation of Smooth Areas of the Greenland Ice Cap, 1953, Part 1," v-vi.
55. Benson, "Stratigraphic Studies in the Snow and Firn of the Greenland Ice Sheet," ii. The logistics are also dealt with in Carl S. Benson, "SIPRE Special Report 17: Resupply of Ice-Cap Expeditions by Airdrop" (US Snow, Ice and Permafrost Research Establishment, 1955); Benson and Ragle, "SIPRE Special Report 18: Project Jello, SIPRE Greenland Expedition 1955, Report on Special Foods Provided by the Quartermaster Food and Container Institute"; Benson, "SIPRE Report 25: Operations and Logistics of Ice-Cap Party Crystal (April 1955)."
56. The aircraft of choice for the drops were the C-54, a military transporter produced by the Douglas Aircraft Company and widely used in non-combat roles including rescue missions and missile tracking and recovery, and the C-119 Flying Boxcar, designed by Fairchild Aircraft, whose entire back end could open to allow supplies to drop out.
57. Benson, "Stratigraphic Studies in the Snow and Firn of the Greenland Ice Sheet," ii.
58. Benson, "SIPRE Special Report 17: Resupply of Ice-Cap Expeditions by Airdrop."
59. Benson, "Oral History Interview with Karen Brewster," 13.

60. Benson, "SIPRE Special Report 17: Resupply of Ice-Cap Expeditions by Airdrop."
61. Benson, "Oral History Interview with Karen Brewster," 17–18.
62. Benson, "SIPRE Special Report 17: Resupply of Ice-Cap Expeditions by Airdrop," 1.
63. Robert Guillard, "Expédition Arctique, Transgroënland 1952: Rapports des Raids et des Transports," n.d. (Expéditions Franco-Américaines au Groënland 1952–1956 Folder, Expéditions Polaires Francaises 1914–2001 Collection (20110210–015), Archives Nationales de France, Fontainebleau); Robert Pommier, "Expédition Arctique, Transgroënland, 1952: Rapport sur les Vêtements et l'Alimentation," n.d. (Expéditions Franco-Americaines au Groënland 1952–1956 Folder, Expéditions Polaires Francaises 1914–2001 Collection (20110210–015), Archives Nationales de France, Fontainebleau); Pommier, *Au-delà de Thule: Sur la Route des Glaces*; Jean-Jacques Holtzscherer, "Expédition Arctique, Transgroënland 1952: Rapport Préliminaire de Sondages Seismiques" (Expéditions Franco-Américaines au Groenland 1952–1956 Folder, Expéditions Polaires Francaises 1914–2001 Collection (20110210–015), Archives Nationales de France, Fontainebleau); Paul-Emile Victor, "Geography of Northeast Greenland, A Report Prepared for SIPRE under Purchase Order 11657, 1955" (CRREL Library, Record ID 33448). As was typical at the time, the contract did not come directly from the military but was mediated by the Stanford Research Institute.
64. See, for example, "Department of State, Washington, 10 February 1966, Note to the Embassy of Denmark Concerning Paul-Emile Victor and EGIG II" (Rigsarkivet, 0030, Grønlandsministeriet Collection, Journalsager 1957–1989, Box 2084, Folder Dansk Fortsættelse af den Internationale Glaciologiske Eksp. 1957/60).
65. During the war, Victor worked for the US Army Air Forces, the direct predecessor of the US Air Force.
66. Benson, "Letter to Paul-Emile Victor, 20 October 1955" (Relations Internationales: Etats-Unis, 1947–1983 Folder, Expéditions Polaires Francaises 1914–2001 Collection (20110210–256), Archives Nationales de France, Fontainebleau).
67. Albert Bauer, "Letter to Henri Bader, 13 August 1955" (Relations Internationales: Etats-Unis, 1947–1983 Folder, Expéditions Polaires Francaises 1914–2001 Collection (20110210–256), Archives Nationales de France, Fontainebleau).
68. Henri Bader, "Letter to Expéditions Polaires Francaises, 28 March 1955" (Relations Internationales: Etats-Unis, 1947–1983 Folder, Expéditions Polaires Francaises 1914–2001 Collection (20110210–256), Archives Nationales de France, Fontainebleau).
69. Paul-Emile Victor, "Letter to Henri Bader, 1 April 1955" (Relations Internationales: Etats-Unis, 1947–1983 Folder, Expéditions Polaires

Francaises 1914–2001 Collection (20110210-256), Archives Nationales de France, Fontainebleau).
70. Benson and Ragle, "SIPRE Special Report 18: Project Jello, SIPRE Greenland Expedition 1955, Report on Special Foods Provided by the Quartermaster Food and Container Institute," 1-21.
71. "Jello: Resumé" (Expéditions Franco-Americaines au Groenland 1952–1956 Folder, Expéditions Polaires Francaises 1914–2001 Collection (20110210-115), Archives Nationales de France, Fontainebleau).
72. Benson, "Stratigraphic Studies in the Snow and Firn of the Greenland Ice Sheet," iv.
73. Benson and Ragle, "SIPRE Special Report 18: Project Jello, SIPRE Greenland Expedition 1955, Report on Special Foods Provided by the Quartermaster Food and Container Institute," 7.
74. Chester C. Langway, "ERDC/CRREL Report TR-08-1: The History of Early Polar Ice Cores" (US Cold Regions Research and Engineering Laboratory, 2008), 6. As well as the overarching glaciological studies which guided the expedition, members of the Jello team also pursued individual research interests. George Wallerstein, officially attached to the project as expedition navigator, conducted a study of the refraction of the sun below the horizon under different situations on the ice sheet (George Wallerstein, "Refraction Observations on the Greenland Icecap," *Navigation* 5, no. 3 (1956)). And the expedition physician, Robert Christie, conducted physiological experiments on the expedition team members to earn credit towards his pathology residency at Dartmouth. To study hemoglobin levels at altitude in cold environments, Christie drew blood from his fellow expedition mates and incubated agar plates using his own body heat by placing them in the pockets of a vest which his wife had designed for the purpose. His work led to a pair of articles in the *New England Journal of Medicine* as well as a trip report in the *Journal of the American Medical Association*, including the first published account of influenza being spread via an airdrop of supplies to the members of an expedition who had been isolated from all outside contact for many weeks (Robert W. Christie, "Bacterial Variations in the Nasopharynx and Skin of Isolated Arctic Scientists," *New England Journal of Medicine* 258 (1958); Robert W. Christie, "Medical Notes on a Greenland Ice Cap Expedition," *Journal of the American Medical Association* 164 (1957)).
75. For Epstein's work in Chicago, see Samuel Epstein et al., "Carbonate-Water Isotopic Temperature Scale," *Geological Society of America Bulletin* 62, no. 4 (1951); Harold C. Urey et al., "Measurement of Paleotemperatures and Temperatures of the Upper Cretaceous of England, Denmark and the Southeastern United States," *Geological Society of America Bulletin* 62, no. 4 (1951). For his work in California, see Samuel Epstein and Toshiko Mayeda, "Variation of O^{18} Content

of Waters from Natural Sources," *Geochimica et Cosmochimica Acta* 4, no. 5 (1953). The key point underlying this work is that summer snow contains relatively more oxygen-18 than winter snow.
76. Benson, "Oral History Interview with Karen Brewster," 11.
77. Langway, "The History of Early Polar Ice Cores."
78. For more on Epstein and Dansgaard's early work, including Epstein's eventual move away from ice core research, see Maiken Lolck, *Klima, Kold Krig og Iskerner* (Aarhus: Aarhus Universitetforlag, 2006).
79. After Germany occupied metropolitan Denmark on April 9, 1940, Denmark retained its own coalition government until 1943, when the government resigned and Germany assumed full control. For Danish political thought in this era, see Poul Villaume, *Allieret Med Forbehold: Danmark, NATO og den Kolde Krig, Et Studie i Dansk Sikkerhedspolitik, 1949–1961* (Copenhagen: Eirene, 1995); "DUPI Vol. 1, Grønland Under den Kolde Krig: Dansk og Amerikansk Sikkerhedspolitik, 1945–68" (Copenhagen: Dansk Udenrigspolitisk Institut, 1997); Geir Lundestad, *America, Scandinavia and the Cold War, 1945–1949* (New York: Columbia University Press, 1980).
80. For French trawlers, see Herbert F. Feaver, "Department of External Affairs Canada Confidential Despatch D-114 from the Canadian Ambassador in Copenhagen, 23 July 1957, Ottawa File No. 6732-40" (LAC, RG85, Denmark and Greenland General File No. 1005-7, Vol. 10, Department of Northern Affairs and National Resources, ATIP Division Interim Box 292-2001099652). For British scientists, see C. James W. Simpson, "The British North Greenland Expedition," *The Geographical Journal* 121, no. 3 (1955). For foreign prospectors, see "Two Million Tons of Uranium Ore on Greenland," translation of article from *Dansk Tidende*, 2 October 1947 (NARA, RG 59, Office of the Secretary of State, Entry 15375, Box 47, Folder Denmark Greenland Nuclear Power Plant). This last subject, and specifically uranium mining in Greenland, is taken up in detail in Henry Nielsen and Henrik Knudsen, "Too Hot to Handle: The Controversial Hunt for Uranium in Greenland in the Early Cold War," *Centaurus* 55 (2013): 319–343.
81. Kjeld Rask Therkilsen, "Greenland and Aviation Serve Each Other," in *Danish Foreign Office Journal* 21 (1956) (LAC, RG85, Denmark and Greenland General File No. 1005-7, Vol. 9, Department of Northern Affairs and National Resources, ATIP Division Interim Box 292-2001099652), 15.
82. For higher level US military concerns about Greenland, see, for example "Proposals with Respect to Greenland, 24 May 1946" (NARA, RG 59, Dec. File 859b.20/5-2446, Box 6515); "Military Requirements for Base Rights, Joint Strategic Plans Committee 684/52, 23 March 1949," Originally Top Secret (NARA, RG 218, Section 36, CCS 360).
83. "DUPI Vol. 1, Grønland Under den Kolde Krig: Dansk og Amerikansk Sikkerhedspolitik, 1945–68," 79.

84. Quoted in Eric S. Einhorn, *National Security and Domestic Politics in Post-War Denmark: Some Principal Issues, 1945–1961* (Odense, Denmark: Odense University Press, 1975), 20. Einhorn provides a thorough analysis of the speech.
85. The results were 80 yes votes, 7 no votes, and 13 abstentions.
86. "The Snowman Project, HQ Atlanta Division, ATC, Westover Field, Massachusetts, 1947" (NARA, RG 341, Air Force: Plans, Project Decimal File 1942–1954, Box 831, SG 581, TS), 43–44. For similar debates in Canada, see Ken Coates et al., eds., *Arctic Front: Defending Canada in the Far North* (Toronto: Thomas Allen & Son, 2008); Lackenbauer and Farish, "The Cold War on Canadian Soil: Militarizing a Northern Environment"; P. Whitney Lackenbauer, ed. *Canada and Arctic Sovereignty and Security: Historical Perspectives* (Calgary: University of Calgary Press—Calgary Papers in Military and Strategic Studies, 2011).
87. "Defense of Greenland 1951, Agreement Between the United States and the Kingdom of Denmark, April 27, 1951" (American Foreign Policy 1950–1955, Basic Documents, Department of State Publication 6446, General Foreign Policy Series 117, Washington DC), Preamble. The negotiation of this agreement is covered in "DUPI Vol. 1, Grønland Under den Kolde Krig: Dansk og Amerikansk Sikkerhedspolitik, 1945–68," 127–70; Nikolaj Petersen, "Negotiating the 1951 Greenland Defense Agreement: Theoretical and Empirical Analyses," *Scandinavian Political Studies* 21 (1998): 1–28; Einhorn, "National Security and Domestic Politics in Post-War Denmark: Some Principal Issues, 1945–1961"; Shelagh D. Grant, *Polar Imperative: A History of Arctic Sovereignty in North America* (Vancouver: Douglas & McIntyre, 2010), 290ff; Carsten Holbrath, *Danish Neutrality: A Study in the Foreign Policy of a Small State* (Oxford: Clarendon, 1991). Views on the agreement are divided: Shelagh Grant argues that Denmark did not fare well in the negotiations, while Nikolaj Petersen takes a more charitable view, emphasizing that Denmark won major concessions on mapping.
88. This process is described in detail in Nikolaj Petersen, "The Politics of US Military Research in Greenland in the Early Cold War," *Centaurus* 55 (2013): 294–318.
89. The Greenland Ministry (which gained its name in 1954) prohibited contact between indigenous Greenlanders and US personnel. See Ibid.
90. Ibid., 9.
91. Aksel Nørvang, "Den Videnskabelige Rådgivers Afsluttende Rapport, 1956" (Rigsarkivet, Udenrigsministeriet Collection, UM 105.F.8, Videnskabelig Rådgiver, Thule, 1956, pp. 116–151).
92. "Referat af Møde i Kommissionen for Videnskabelige Undersøgelser i Grønland, onsdag den 3. maj 1961 kl. 15 i Arktisk Institut" (Rigsarkivet, 0030, Grønlandsministeriet Collection, Journalsager 1957–1989, Box

2084, Folder Dansk Fortsættelse af den Internationale Glaciologiske Eksp. 1957/60), 9.
93. Hans Christiansen, "Notits: Amerikansk Videnskabelig Virksomhed i Grønland," 16 June 1958 (Rigsarkivet, Udenrigsministeriet Collection, 105.F.9.A, US Aktivitet Udenfor Forsvarsområderne, 1957–1963), 2.
94. Robert Coe, "Despatch to the Department of State, 611.59/5–2857, 28 May 1957, Confidential, Drafted by James W. Gantenbein" (Foreign Relations of the United States, 1955–1957: Volume XXVII, Western Europe and Canada, Department of State, Central Files, Document 180).
95. Knudsen, "Cold War, Ionospheric Research in Greenland, and the Politics of Rockets: A Study of the Ill-Fated Operation PCA 68." Also see Thorsten Borring Olesen, "Tango for Thule: The Dilemmas and Limits to the 'Neither Confirm nor Deny Doctrine' in the Danish-American Relationship, 1957–1968," *Journal of Cold War Studies* 13 (2011).
96. For Project Iceworm, see Erik D. Weiss, "Cold War Under the Ice: The Army's Bid for a Long-Range Nuclear Role, 1959–1963," *Journal of Cold War Studies* 3 (2001); Nikolaj Petersen, "The Iceman That Never Came: 'Project Iceworm', the Search for a NATO Deterrent, and Denmark, 1960–62," *Scandinavian Journal of History* 33 (2008).
97. Lucius D. Battle, "Confidential Letter from the US Embassy in Copenhagen to Birger Kronmann, Danish Ministry of Foreign Affairs, 14 July 1954" (Rigsarkivet, Udenrigsministeriet Collection, 105.F.9.A, US Aktivitet Udenfor Forsvarsområderne, 1952–1957); Benson, "Oral History Interview with Karen Brewster," 8.
98. There is no mention of the project in the permissions folder of the Danish Foreign Ministry archive (Udenrigsministeriet Arkiv, Copenhagen, 105.F.9.a: US Aktivitet Udenfor Forsvarsområderne, 1952–1957). See, for example, Winston Butscher, "Scope of 1955 TC Arctic Activities in Greenland, Doc. No. 9-98-07-002" (Rigsarkivet, Udenrigsministeriet Collection, 105.F.9.A, US Aktivitet Udenfor Forsvarsområderne, 1952–1957). It is not clear why the US did not inform the Danes about the traverse.
99. Benson, "Oral History Interview with Karen Brewster," 8.
100. Heymann et al., "Exploring Greenland: Science and Technology in Cold War Settings."
101. Lloyd H. Cornett and Mildred W. Johnson, *A Handbook of Aerospace Defense Organization 1946–1980* (Colorado: Office of History, Aerospace Defense Center, Peterson Air Force Base, 1980).
102. Fristrup, *The Greenland Ice Cap*, 146–47.
103. This nuclear reactor was installed at Camp Century; see Lawrence H. Suid, *The Army's Nuclear Power Program: The Evolution of a Support Agency* (Westport, Connecticut: Greenwood Press, 1990).
104. Benson, "Oral History Interview with Karen Brewster," 7, 29.

4 It Has Completely Changed

1. "Jökull," 1951 (Yearbook of the Glaciological Society of Iceland, Vol. 1); Richard Haefeli, "The Development of Snow and Glacier Research in Switzerland," *Journal of Glaciology* 1, no. 4 (1948): 193; E. Barillon, "Correspondance," *Journal of Glaciology* 1, no. 6 (1949): 337.
2. Robert MacDonald, "Challenges and Accomplishments: A Celebration of the Arctic Institute of North America," *Arctic* 58, no. 4 (2005).
3. Robert F. Legget, "Canadian Snow Conference," *Journal of Glaciology* 1, no. 3 (1948); "Glaciological Conference of the American Geographical Society," *Journal of Glaciology* 1, no. 6 (1949).
4. "Conference on Glaciological Research Under the Auspices of the Arctic Institute of North America and the American Geographical Society, New York City" (American Geographical Society Report, 1949).
5. G. Seligman, "Meeting of the International Commission on Snow and Glaciers, Oslo," *Journal of Glaciology* 1, no. 5 (1948); Uwe Radok, "The International Commission on Snow and Ice and Its Precursors, 1894–1994," *Hydrological Sciences Journal* 42, no. 2 (1997).
6. "Réunion de Constitution de l'EGIG, Grindelwald," April 1956 (Organisation EGIG: Projets, Réunions Préliminaires 1954–1957 Folder, Expéditions Polaires Francaises 1914–2001 Collection (20110210–117), Archives Nationales de France, Fontainebleau); Richard Haefeli, "Projet de Programme Scientifique de l'EGIG, 3 September 1955" (Programmes Scientifiques EGIG I Folder, Expéditions Polaires Francaises 1914–2001 Collection (20110210–127), Archives Nationales de France, Fontainebleau).
7. The Federal Republic of Germany's participation in EGIG grew out of personal connections between French scientists Paul-Emile Victor and Albert Bauer and German seismologist Bernard Brockamp. Through the exchange of letters and scientific papers in the early 1950s, Victor and Bauer became aware of Brockamp's desire to reintegrate Germany into polar research, and soon came to champion that desire. Bauer's support for this cause is especially striking given his war experience: from his post as a French artillery lieutenant, he was captured and held near the Austrian-Czech border for five years—during which, as an Alsatian, he faced constant pressure to declare allegiance to the Axis powers. Although still sensitive from occupation during the war, and especially from the continued problems facing the Danish minority in Schleswig-Holstein, Denmark granted permission for its old enemy and new NATO ally to join EGIG—but refused to allow German aircraft to take part (for Schleswig-Holstein, see Eric S. Einhorn, *National Security and Domestic Politics in Post-War Denmark: Some Principal Issues, 1945–1961* (Odense, Denmark: Odense University Press, 1975), 36–37). The Danes' unease softened by 1959, when Børge Fristrup commented that "it is very gratifying that there has never been a conflict

of national character or national differences, despite the fact that this [that is, EGIG] has involved people who participated in the German occupation of Versailles during the last World War, and Frenchmen who were held in German POW camps, or who participated in the French Resistance" (Børge Fristrup, "Rapport Vedrørende: Expédition Glaciologique Internationale au Groënland," November 1959 (Rigsarkivet, 0030, Grønlandsministeriet Collection, Journalsager 1957–1989, Box 2083, Folder Internationale Glaciologiske Eksp., Victor)). By 1964, the Danish Foreign Office and Ministry of Defense took a further step by allowing *Luftwaffe* airplanes to support scientific research in Greenland, "on the condition, however, that the German participation be limited to a few unarmed planes" (Einar Andersen, "Letter to Walther Hofmann, 31 August 1964" (Rigsarkivet, 0030, Grønlandsministeriet Collection, Journalsager 1957–1989, Box 2084, Folder Dansk Fortsættelse af den Internationale Glaciologiske Eksp. 1957/60). Also see "Letter to Paul-Emile Victor, 9 September 1964" (Rigsarkivet, 0030, Grønlandsministeriet Collection, Journalsager 1957–1989, Box 2084, Folder Dansk Fortsættelse af den Internationale Glaciologiske Eksp. 1957/60).). Notably, EGIG represents the first return of Germans to the polar world after World War II (see Christian Kehrt, "Eternal Ice in the Cold War: The Polar Regions in Spatial and Environmental Perspective, 1957–1991," Paper presented at the *Cold War Science, Colonial Politics and National Identity in the Arctic Workshop*, Aarhus, December 2010).
8. André Renaud, "La Participation de la Suisse à l'Expédition Glaciologique Internationale (EGIG) de 1957 à 1961," *Revue Internationale de l'Horlogerie* (1958): U1.
9. Ibid.: U2.
10. Through this period, Victor juggled his EGIG work with his commitments in Terre Adélie, where EPF was in the process of establishing a permanent base.
11. First quote: Richard Finsterwalder, "Expédition Glaciologique Internationale au Groënland 1957–60 (EGIG)," *Journal of Glaciology* 26, no. 3 (1959): 543. Emphasis in original. Second quote: "Réunion de Constitution de l'EGIG, Grindelwald," April 1956 (Organisation EGIG: Projets, Réunions Préliminaires 1954–1957 Folder, Expéditions Polaires Francaises 1914–2001 Collection (20110210–117), Archives Nationales de France, Fontainebleau).
12. Paul-Emile Victor, "Letter to Richard Haefeli, 1 September 1955" (Organisation EGIG: Projets, Réunions Préliminaires 1954–1957 Folder, Expéditions Polaires Francaises 1914–2001 Collection (20110210–117), Archives Nationales de France, Fontainebleau); "Rapport Général d'Activités Arctiques et Antarctiques Depuis 1947," December 1970 (Expéditions Franco-Americaines au Groënland 1952–1956 Folder, Expéditions Polaires Francaises 1914–2001 Collection (20110210–015), Archives Nationales de France, Fontainebleau), 10.

13. See, for example, "Note Verbale No. 262, Embassy of France in the USA to the State Department, Washington, DC, 15 May 1958" (Rigsarkivet, 0030, Grønlandsministeriet Collection, Journalsager 1957–1989, Box 2083, Folder Internationale Glaciologiske Eksp., Victor); "Note Verbale No. 10, Embassy of France in Copenhagen to the Danish Ministry of Foreign Affairs, 16 January 1957" (Rigsarkivet, 0030, Grønlandsministeriet Collection, Journalsager 1957–1989, Box 2083, Folder Internationale Glaciologiske Eksp., Victor); Paul-Emile Victor, "Letter to P. Benoist, Direction d'Amérique, Ministère des Affaires Etrangères, Paris, 3 January 1957" (Rigsarkivet, 0030, Grønlandsministeriet Collection, Journalsager 1957–1989, Box 2083, Folder Internationale Glaciologiske Eksp., Victor).
14. Willi Dansgaard, *Frozen Annals: Greenland Ice Cap Research* (Copenhagen: Niels Bohr Institute, 2005), 32.
15. Albert Bauer, "Letter to Richard Haefeli, 21 January 1956" (Organisation EGIG: Projets, Réunions Préliminaires 1954–1957 Folder, Expéditions Polaires Francaises 1914–2001 Collection (20110210–117), Archives Nationales de France, Fontainebleau); Pierre Pflimlin, "Letter to Paul-Emile Victor, 11 July 1955" (Organisation EGIG: Projets, Réunions Préliminaires 1954–1957 Folder, Expéditions Polaires Francaises 1914–2001 Collection (20110210–117), Archives Nationales de France, Fontainebleau); "Ministère de l'Education Nationale, Paris to Monsieur Vaugelade, 18 January 1956" (Budget EGIG: Documents Divers 1955–1970 Folder, Expéditions Polaires Francaises 1914–2001 Collection (20110210–122), Archives Nationales de France, Fontainebleau).
16. For de Gaulle and the National Center for Scientific Research, see Antoine Prost, *Les Réformes du CNRS: 1959–1960* (Paris: Cahiers Pour l'Histoire du CNRS, 1990).
17. Albert Bauer, "Letter to Monsieur de Directeur, Institut Géographique National, Paris, 29 August 1958" (Programmes Scientifiques EGIG I Folder, Expéditions Polaires Francaises 1914–2001 Collection (20110210–127), Archives Nationales de France, Fontainebleau); "Rapport Général d'Activités Arctiques et Antarctiques Depuis 1947," December 1970 (Expéditions Franco-Americaines au Groënland 1952–1956 Folder, Expéditions Polaires Francaises 1914–2001 Collection (20110210–015), Archives Nationales de France, Fontainebleau), 10. The French Armée de l'Air provided air support from 1957–1960, after which growing problems in Algeria meant that the French military could no longer guarantee support for EGIG. EGIG then turned to the US for help in evacuating men and equipment from the ice sheet at the end of the expedition. See Paul-Emile Victor, "Letter to Colonel Chappell, 3 February 1960" (Relations Internationales: Etats-Unis, 1947–1983 Folder, Expéditions Polaires Francaises 1914–2001 Collection (20110210–256), Archives Nationales de France, Fontainebleau).

18. "EGIG Doc. 8, Compte-Rendu de la 3eme Séance de la Réunion de Mise au Point de l'EGIG à Grindelwald," 5 April 1956 (Organisation EGIG: Projets, Réunions Préliminaires 1954–1957 Folder, Expéditions Polaires Francaises 1914–2001 Collection (20110210–117), Archives Nationales de France, Fontainebleau); Paul-Emile Victor, "Note pour le Secrétaire General EGIG, Objet: Organisation de l'EGIG," 12 February 1957 (Organisation EGIG: Projets, Réunions Préliminaires 1954–1957 Folder, Expéditions Polaires Francaises 1914–2001 Collection (20110210–117), Archives Nationales de France, Fontainebleau); Renaud, "La Participation de la Suisse à l'Expédition Glaciologique Internationale (EGIG) de 1957 à 1961," U4.
19. Børge Fristrup, "Rapport over den Internationale Glaciologiske Expeditionen," 3 September 1959 (Rigsarkivet, 0030, Grønlandsministeriet Collection, Journalsager 1957–1989, Box 2083, Folder Internationale Glaciologiske Eksp., Victor).
20. Dansgaard, *Frozen Annals: Greenland Ice Cap Research*, 30.
21. "Ekstrakt af Referat af Møde den 29. april 1958, Punkt 1, De Geofysiske Undersøgelser i Grønland, Herunder den Internationale Glaciologiske Ekspedition" (Rigsarkivet, 0030, Grønlandsministeriet Collection, Journalsager 1957–1989, Box 2083, Folder Internationale Glaciologiske Eksp., Victor).
22. Paul-Emile Victor, "Letter to M. Martinet, Ministère Des Finances, 2 November 1955" (Budget EGIG: Documents Divers 1955–1970 Folder, Expéditions Polaires Francaises 1914–2001 Collection (20110210–122), Archives Nationales de France, Fontainebleau).
23. Henri Bader, "Letter to Albert Bauer," n.d., très confidentiel (Relations Internationales: Etats-Unis, 1947–1983 Folder, Expéditions Polaires Francaises 1914–2001 Collection (20110210–256), Archives Nationales de France, Fontainebleau).
24. "Minutes, 4th Meeting, Technical Panel on Glaciology, Pasadena, California, 29–30 August 1955" (NARA, RG 27, Entry No. 3, Geophysical Year 1953–1960, 130/16/9/2 Box 3, Glaciology). For Bader, the IGY, and US ice core work, see Janet Martin-Nielsen, "'The Deepest and Most Rewarding Hole Ever Drilled': Ice Cores and the Cold War in Greenland," *Annals of Science* 70 (2013): 58–59.
25. There is a huge literature on the IGY; for an overview, see Roger D. Launius, "Toward the Poles: A Historiography of Scientific Exploration During the International Polar Years and the International Geophysical Year," in *Globalizing Polar Science: Reconsidering the International Polar and Geophysical Years*, ed. James R. Fleming and Roger D. Launius (New York: Palgrave Macmillan, 2010), 66.
26. The Geophysical Institute at the University of Alaska Fairbanks and the Roger G. Barry Archives at the National Snow and Ice Data Center in Boulder, CO, host collections of Russian-language documents from the Soviet IGY program; see, for example, "Russian-language IGY

Information Bulletins," USSR Academy of Science, Moscow, 1957–1964 (NSIDC, ARC Archives Catalog ID 0B324306).
27. See, for example, "IGY General Report no. 20, November 1963" (National Academies of Science, Washington DC). For the World Data Centers, see Fae L. Korsmo, "The Origins and Principles of the World Data Center System," *Data Science Journal* 8 (2010).
28. "Météorologie Antarctique: l'Année Géophysique Internationale en Terre Adélie," in *Revue Trimestrielle de Météorologie et de Physique du Globe, Annuaire de la Société Météorologique de France*, January 1960 (NSIDC, ARC Archives Catalog ID 0B3242E5). The relationship between Expéditions Polaires Françaises and the IGY is complex; see John Hanessian, "Letter to Richard Nolte, 12 April 1960" (Institute of Current World Affairs Archives, Document 4/12/1960/JH-14); Leonard A. LeSchack, "The French Polar Effort and the Expéditions Polaires Francaises," *Arctic* 17 (1964).
29. "EGIG Doc. 8, Compte-Rendu de la 3eme Séance de la Réunion de Mise au Point de l'EGIG à Grindelwald," 5 April 1956 (Organisation EGIG: Projets, Réunions Préliminaires 1954–1957 Folder, Expéditions Polaires Francaises 1914–2001 Collection (20110210–117), Archives Nationales de France, Fontainebleau); "EGIG: Campagne d'Eté 1959, Compte-Rendu Opérationnel Préliminaire pour le Comité de Direction d'Hambourg," 26–29 October 1959 (EGIG I: Réunions du Comité de Direction, 1955–1959 Folder, Expéditions Polaires Francaises 1914–2001 Collection (20110210–118), Archives Nationales de France, Fontainebleau). For a detailed description of the campaign, see Thierry Fournier, "Chapitre VIII: Paul-Emile Victor et l'Expédition Glaciologique Internationale au Groënland, 1955–1960" (Unpublished PhD thesis chapter, Ecole Nationale des Chartes, France, 2012).
30. "EGIG Doc. 8, Compte-Rendu de la 3eme Séance de la Réunion de Mise au Point de l'EGIG à Grindelwald," 5 April 1956 (Organisation EGIG: Projets, Réunions Préliminaires 1954–1957 Folder, Expéditions Polaires Francaises 1914–2001 Collection (20110210–117), Archives Nationales de France, Fontainebleau); Byron B. Webb, "Special Staff Meeting of 6 May 1958, 1000 Hours, Headquarters, 4084th Air Base Group, United States Air Force, APO 121, New York" (Rigsarkivet, 0030, Grønlandsministeriet Collection, Journalsager 1957–1989, Box 2083, Folder Internationale Glaciologiske Eksp., Victor).
31. Albert L. Washburn, "Letter to Albert Bauer, 31 October 1955" (Programmes Scientifiques EGIG I Folder, Expéditions Polaires Francaises 1914–2001 Collection (20110210–127), Archives Nationales de France, Fontainebleau).
32. Paul-Emile Victor, "Letter to Jack Crowell, 10 December 1959"; "Jello Cache: Sondrestrom" (Relations Internationales: Etats-Unis, 1947–1983 Folder, Expéditions Polaires Francaises 1914–2001 Collection (20110210–256), Archives Nationales de France, Fontainebleau);

"Inventory of Equipment Cached by 1955 Expedition Jello at 69.44'N, 48.03'W" (Expéditions Franco-Americaines au Groënland 1952–1956 Folder, Expéditions Polaires Francaises 1914–2001 Collection (20110210–115), Archives Nationales de France, Fontainebleau).
33. "Jello Cache: Sondrestrom" (Relations Internationales: Etats-Unis, 1947–1983 Folder, Expéditions Polaires Francaises 1914–2001 Collection (20110210–256), Archives Nationales de France, Fontainebleau).
34. "Réunion de Constitution de l'EGIG, Grindelwald," April 1956 (Organisation EGIG: Projets, Réunions Préliminaires 1954–1957 Folder, Expéditions Polaires Francaises 1914–2001 Collection (20110210–117), Archives Nationales de France, Fontainebleau).
35. Eske Brun, "Letter to Albert Bauer, 22 March 1961" (Rigsarkivet, 0030, Grønlandsministeriet Collection, Journalsager 1957–1989, Box 2083, Folder Børge Fristrup, 1420–21–02).
36. "Continuous Observations at the Greenland Ice Cap Station" (Organisation EGIG: Projets, Réunions Préliminaires 1954–1957 Folder, Expéditions Polaires Francaises 1914–2001 Collection (20110210–117), Archives Nationales de France, Fontainebleau).
37. Fristrup, "Rapport Vedrørende: Expédition Glaciologique Internationale au Groënland," November 1959 (Rigsarkivet, 0030, Grønlandsministeriet Collection, Journalsager 1957–1989, Box 2083, Folder Internationale Glaciologiske Eksp., Victor).
38. "Groupes Scientifiques, EGIG" (Programmes Scientifiques EGIG I Folder, Expéditions Polaires Francaises 1914–2001 Collection (20110210–127), Archives Nationales de France, Fontainebleau); Hermann Mälzer, *Das Nivellement Über das Grönländische Inlandeis der Internationalen Glaziologischen Grönland-Expedition 1959* (Copenhagen: Meddelelser om Grønland, 1964).
39. Marcel de Quervain, "Internationale Glaziologische Groenlandexpedition, Groupe Schneekunde, Schweiz: Programmvorschläge zur Schneekunde" (Programmes Scientifiques EGIG I Folder, Expéditions Polaires Francaises 1914–2001 Collection (20110210–127), Archives Nationales de France, Fontainebleau); Marcel de Quervain, *Schneekundliche Arbeiten der Internationalen Glaziologischen Grönlandexpedition: Nivologie* (Copenhagen: Meddelelser om Grønland, 1969); Albert Bauer, "EGIG Campagne de 1960: Projet des Recherches à Effectuer" (Programmes Scientifiques EGIG I Folder, Expéditions Polaires Francaises 1914–2001 Collection (20110210–127), Archives Nationales de France, Fontainebleau); Fristrup, "Rapport Vedrørende: Expédition Glaciologique Internationale au Groënland," November 1959 (Rigsarkivet, 0030, Grønlandsministeriet Collection, Journalsager 1957–1989, Box 2083, Folder Internationale Glaciologiske Eksp., Victor).
40. Matthias Heymann, "The Evolution of Climate Ideas and Knowledge," *Wiley Interdisciplinary Reviews: Climate Change* 1, no. 4

(2010): 588–89. Also see Matthias Heymann, "Klimakonstruktionen: Von der Klassichen Klimatologie zur Klimaforschung," *NTM Journal of History of Sciences, Technology and Medicine* 17 (2009).
41. Hubert Horace Lamb, "Britain's Climate in the Past (Lecture given to the British Association for the Advancement of Science at Southampton on September 2, 1964)," in *The Changing Climate: Selected Papers*, ed. Hubert Horace Lamb (London: Methuen & Co., 1966), 171.
42. Ibid., 171–72.
43. The leading work on the history of global warming is Spencer Weart, *The Discovery of Global Warming* (Cambridge, MA: Harvard University Press, 2008). For Ahlmann, see Sverker Sörlin, "Narratives and Counter Narratives of Climate Change: North Atlantic Glaciology and Meteorology, c. 1930–1955," *Journal of Historical Geography* 35 (2009); Sverker Sörlin, "The Global Warming That Did Not Happen: Historicizing Glaciology and Climate Change," in *Nature's End: Environment and History*, ed. Sverker Sörlin and Paul Warde (London: Palgrave Macmillan, 2009).
44. See, for example, Renaud, "La Participation de la Suisse à l'Expédition Glaciologique Internationale (EGIG) de 1957 à 1961."
45. "Groupes Scientifiques, EGIG" (Programmes Scientifiques EGIG I Folder, Expéditions Polaires Francaises 1914–2001 Collection (20110210–127), Archives Nationales de France, Fontainebleau); André Renaud, *Etudes Physiques et Chimiques sur la Glace de l'Indlandsis du Groënland 1959* (Copenhagen: Meddelelser om Grønland, 1969).
46. Renaud, *Etudes Physiques et Chimiques sur la Glace de l'Indlandsis du Groënland 1959*, 76ff; Renaud, "La Participation de la Suisse à l'Expédition Glaciologique Internationale (EGIG) de 1957 à 1961"; André Renaud, "EGIG: Programme de la Commission des Glaciers de la Société Helvétique des Sciences Naturelles (B. Physique et Chimie de la Glace)" (Programmes Scientifiques EGIG I Folder, Expéditions Polaires Francaises 1914–2001 Collection (20110210–127), Archives Nationales de France, Fontainebleau).
47. Renaud, *Etudes Physiques et Chimiques sur la Glace de l'Indlandsis du Groënland 1959*, 88ff.
48. See, for example, V.N. Nijampurkar and Henrik B. Clausen, "A Century Old Record of Lead-210 Fallout on the Greenland Ice Sheet," *Tellus* 42B (1990); Toshitaka Suzuki et al., "Determination of Lead-210 in an Ice Core from an Arctic Glacier by an Alpha-Ray Measurement," *Analytical Sciences* 12 (1996); Jack E. Dibb and Henrik B. Clausen, "A 200-Year Pb-210 Record from Greenland," *Journal of Geophysical Research* 102, no. D4 (1997).
49. Dansgaard, *Frozen Annals: Greenland Ice Cap Research*, 31–32.
50. Renaud, *Etudes Physiques et Chimiques sur la Glace de l'Indlandsis du Groënland 1959*, 70.

51. Dansgaard, *Frozen Annals: Greenland Ice Cap Research*, 37. Also see Willi Dansgaard and Anker Weidick, "Klimaforværring i Grønland?," *Tidsskriftet Grønland* (November 1965).
52. Douglas Martin, "Willi Dansgaard Dies at 88; Read Climates in Old Ice," *The New York Times* (January 29, 2011), D7.
53. Janet Martin-Nielsen, "The Other Cold War: The United States and Greenland's Ice Sheet Environment, 1948–1966," *Journal of Historical Geography* 38 (2012).
54. "Note pour M. Paul-Emile Victor, Paris, 25 January 1956" (Organisation EGIG: Projets, Réunions Préliminaires 1954–1957 Folder, Expéditions Polaires Francaises 1914–2001 Collection (20110210–117), Archives Nationales de France, Fontainebleau); Gaston Rouillon, "Compte-Rendu de l'Entrevue de G. Rouillon à l'Ambassade du Danemark le 22 Decembre 1955" (Relations avec le Danemark, 1947–1991 Folder, Expéditions Polaires Francaises 1914–2001 Collection (20110210–255), Archives Nationales de France, Fontainebleau); Børge Fristrup, "Noter til Mødet Med EGIG D. 8–9/5–1961" (Rigsarkivet, 0030, Grønlandsministeriet Collection, Journalsager 1957–1989, Box 2083, Folder Paul-Emile Victor, Ekspeditioner, 1420–21–01).
55. Denmark's scientific nationalism is further discussed in the epilogue.
56. Rouillon, "Compte-Rendu de l'Entrevue de G. Rouillon à l'Ambassade du Danemark le 22 Decembre 1955" (Relations avec le Danemark, 1947–1991 Folder, Expéditions Polaires Francaises 1914–2001 Collection (20110210–255), Archives Nationales de France, Fontainebleau).
57. Eske Brun, "Letter to Paul-Emile Victor, 20 June 1956" (Rigsarkivet, 0030, Grønlandsministeriet Collection, Journalsager 1957–1989, Box 2083, Folder Internationale Glaciologiske Eksp., Victor).
58. Nils Svenningsen, "Letter to Eske Brun, 13 June 1956" (Rigsarkivet, 0030, Grønlandsministeriet Collection, Journalsager 1957–1989, Box 2083, Folder Internationale Glaciologiske Eksp., Victor).
59. Paul-Emile Victor, "Letter to Eske Brun, 3 January 1957" (Relations avec le Danemark, 1947–1991 Folder, Expéditions Polaires Francaises 1914–2001 Collection (20110210–255), Archives Nationales de France, Fontainebleau); "Letter to Eske Brun, 28 September 1955" (Relations avec le Danemark, 1947–1991 Folder, Expéditions Polaires Francaises 1914–2001 Collection (20110210–255), Archives Nationales de France, Fontainebleau); "Letter to Ambassade de Danemark, Paris, 30 August 1955" (Relations avec le Danemark, 1947–1991 Folder, Expéditions Polaires Francaises 1914–2001 Collection (20110210–255), Archives Nationales de France, Fontainebleau).
60. Albert Bauer, "Letter to Eske Brun, 5 January 1956" (Relations avec le Danemark, 1947–1991 Folder, Expéditions Polaires Francaises 1914–2001 Collection (20110210–255), Archives Nationales de France, Fontainebleau); "Letter to Eske Brun, 9 March 1956"

(Relations avec le Danemark, 1947–1991 Folder, Expéditions Polaires Francaises 1914–2001 Collection (20110210–255), Archives Nationales de France, Fontainebleau); Fristrup, "Noter til Mødet Med EGIG D. 8–9/5–1961" (Rigsarkivet, 0030, Grønlandsministeriet Collection, Journalsager 1957–1989, Box 2083, Folder Paul-Emile Victor, Ekspeditioner, 1420–21–01).

61. Ejnar Wærum, "Letter to Paul-Emile Victor, 27 March 1956" (Relations avec le Danemark, 1947–1991 Folder, Expéditions Polaires Francaises 1914–2001 Collection (20110210–255), Archives Nationales de France, Fontainebleau); Paul-Emile Victor, "Letter to Colonel Jorgen Helk, 3 May 1956" (Relations avec le Danemark, 1947–1991 Folder, Expéditions Polaires Francaises 1914–2001 Collection (20110210–255), Archives Nationales de France, Fontainebleau).

62. Paul-Emile Victor, "Letter to Monsieur le Directeur de l'Institut de Géodesie du Danemark, 6 December 1958" (Relations avec le Danemark, 1947–1991 Folder, Expéditions Polaires Francaises 1914–2001 Collection (20110210–255), Archives Nationales de France, Fontainebleau); Fristrup, "Rapport Vedrørende: Expédition Glaciologique Internationale au Groënland," November 1959 (Rigsarkivet, 0030, Grønlandsministeriet Collection, Journalsager 1957–1989, Box 2083, Folder Internationale Glaciologiske Eksp., Victor).

63. Svenningsen, "Letter to Eske Brun, 13 June 1956" (Rigsarkivet, 0030, Grønlandsministeriet Collection, Journalsager 1957–1989, Box 2083, Folder Internationale Glaciologiske Eksp., Victor). Also see "Note Verbale P.J.V. No 8 X.135, Ambassade de France, Copenhague," 25 February 1957 (Relations avec le Danemark, 1947–1991 Folder, Expéditions Polaires Francaises 1914–2001 Collection (20110210–255), Archives Nationales de France, Fontainebleau).

64. Marianne Krogh Andersen, "Kongen af Grønland," *Weekendavisen* (February 4, 2011), 8–9. For Brun, also see Jens Heinrich, *Eske Brun og det Moderne Grønlands Tilblivelse 1932–1964* (PhD Thesis, Ilisimatusarfik/Grønlands Universitet, 2010). Brun's memoirs also make an interesting read; see Eske Brun, *Mit Grønlandsliv: Erindringer af Eske Brun* (Denmark: Gyldendal, 1985).

65. For the Thule relocation, see Jens Brøsted and Mads Fægteborg, *Thule: Fangerfolk og Militæranlæg* (Copenhagen: Akademisk Forlag, 1985).

66. Brun, "Letter to Paul-Emile Victor, 20 June 1956" (Rigsarkivet, 0030, Grønlandsministeriet Collection, Journalsager 1957–1989, Box 2083, Folder Internationale Glaciologiske Eksp., Victor).

67. "Ekstrakt af Referat af Møde den 29. april 1958, Punkt 1, De Geofysiske Undersøgelser i Grønland, Herunder den Internationale Glaciologiske Ekspedition" (Rigsarkivet, 0030, Grønlandsministeriet Collection, Journalsager 1957–1989, Box 2083, Folder Internationale Glaciologiske Eksp., Victor).

68. Ibid.
69. Fristrup, "Noter til Mødet Med EGIG D. 8–9/5–1961" (Rigsarkivet, 0030, Grønlandsministeriet Collection, Journalsager 1957–1989, Box 2083, Folder Paul-Emile Victor, Ekspeditioner, 1420-21-01). Also see "Aide-Mémoire, EGIG II, 1961" (Rigsarkivet, 0030, Grønlandsministeriet Collection, Journalsager 1957–1989, Box 2084, Folder Dansk Fortsættelse af den Internationale Glaciologiske Eksp. 1957/60), 4.
70. "Meeting in Copenhagen, 8–9 May 1961" (Rigsarkivet, 0030, Grønlandsministeriet Collection, Journalsager 1957–1989, Box 2083, Folder Paul-Emile Victor, Ekspeditioner, 1420-21-01).
71. Ibid.
72. Ibid.
73. See, for example, "La Piste Blanche: Armor Films Présentera les Trois Derniers Courts Métrages de Mario Marret, Réalisé au Groenland au Cours de la Campagne d'Eté de l'Expédition Glaciologique Internationale au Groënland, Paris," 7 March 1960 (Rigsarkivet, 0030, Grønlandsministeriet Collection, Journalsager 1957–1989, Box 2084, Folder Dansk Fortsættelse af den Internationale Glaciologiske Eksp. 1957/60).
74. John Dille, "Mapping the Earth," *Life Magazine*, 1965, 132.
75. Floyd W. Hough, "Note to Acting Executive Director, Army Map Service, April 2 1947" (NARA, RG 77, Army Map Service Box 2–3, Folder 914).
76. Dille, "Mapping the Earth," 129.
77. "Long-Term Scientific Studies for the Standing Group North Atlantic Treaty Organization by Working Group III on Geophysics (Von Karman Committee)," March 1961, Originally Secret (NATO, VKC-EX1-GPIII).
78. Ibid.
79. Dille, "Mapping the Earth," 138. For US geodesy and cartography during the Cold War, see John Cloud, "American Cartographic Transformations During the Cold War," *Cartography and Geographic Information Science* 29, no. 3 (2002); Deborah J. Warner, "From Tallahassee to Timbuktu: Cold War Efforts to Measure Intercontinental Distances," *Historical Studies in the Physical and Biological Sciences* 30 (2000). For the Soviet side of geodetic mapping in the Cold War, a project of staggering scope and accuracy, see John Davies, "Uncle Joe Knew Where You Lived, Part 1," *Sheetlines: Journal of the Charles Close Society for the Study of Ordnance Survey Maps* 72, April (2005); John Davies, "Uncle Joe Knew Where You Lived, Part 2," *Sheetlines: Journal of the Charles Close Society for the Study of Ordnance Survey Maps* 73, August (2005).
80. The US Army (namely, the Army Map Service) also played a critical role in postwar geodesy. For the colorful world of US inter-service rivalries in Cold War geodesy, see Deborah J. Warner, "Political

Geodesy: The Army, the Air Force and the World Geodetic System of 1960," *Annals of Science* 59 (2002).
81. Dille, "Mapping the Earth," 126.
82. Clair E. Ewing quoted in Veikko A. Heiskanen, "Symposium: New Era of Geodesy," *Chronique Internationale* 3 (1954): 67.
83. For SHORAN's early history, see H.G. Sennert, "The History of SHORAN," *Photogrammetric Engineering* 12, no. 4 (1946); Stuart William Seeley, "SHORAN: A Precision Five Hundred Meter Yardstick," *Proceedings of the American Philosophical Society* 105, no. 4 (1961); Alex Green and Jesse Gordon, "Slide Rules and WWII Bombing: A Personal History," *Journal of the Oughtred Society* 21, no. 2 (2012).
84. Carl I. Aslakson and Donald A. Rice, "The Use of SHORAN in Geodetic Control," *American Geophysical Union Transactions* 27 (1946); Carl I. Aslakson, *The Principles of SHORAN Mapping* (Philadelphia, PA: Aero Service Corporation, Photogrammetic Engineers, 1957). Also see Albert E. Theberge, "The Making of an Earth Measurer," *Hydro International* 13, no. 7 (2009).
85. John E.R. Ross, "Geodetic Observations in the Canadian Arctic," *Arctic* 7 (1954); Angus C. Hamilton, "Geodetic Survey of Northern Canada by SHORAN Trilateration," *Polar Record* 9, no. 61 (1959).
86. Seeley, "SHORAN: A Precision Five Hundred Meter Yardstick," 449.
87. J. Kermit Walls, "The RC-130A Aircraft: A New World Mapping System," *Photogrammetric Engineering* (June 1960); Paul W. Jordan, "HIRAN Instrumental Developments," *Journal of Geophysical Research* 65, no. 2 (1960).
88. In the literature, HIRAN is also described as a contraction of HIgh frequency RAnging and Navigation.
89. Veikko A. Heiskanen, "New Era of Geodesy," *Science* 121, no. 3133 (1955): 50; Warner, "From Tallahassee to Timbuktu: Cold War Efforts to Measure Intercontinental Distances," 396; Lee R. Williams, "Final Report, Project 53-AFS-1, Phase 1" (Headquarters, 55th Strategic Reconnaissance Wing Detachment #2, US Air Force, 1953); "APCS: An Exceptional Unit," *The MATS Flyer* 11 (1964); James Raymond Smith, *Introduction to Geodesy: The History and Concepts of Modern Geodesy* (New York: John Wiley & Sons, 1997).
90. For US approaches to clandestine geodetic knowledge from behind the Curtains, see John Cloud, "Hidden in Plain Sight: The Corona Reconnaissance Satellite Program and the Cold War Convergence of the Earth Sciences," *Annals of Science* 58 (2001); Curtis Peebles, *The Corona Project: America's First Spy Satellites* (Annapolis, MD: Naval Institute Press, 1997).
91. Cloud, "American Cartographic Transformations During the Cold War," 266. For WGS 1960, see "Defense Mapping Agency Technical Report 80–003: Geodesy for the Layman" (Washington, DC: Defense Technical Information Center, 1984), Chapter VIII; Deborah J.

NOTES

Warner, "Political Geodesy: The Army, the Air Force and the World Geodetic System of 1960," *Annals of Science* 59 (2002): 363–89.

92. Winston Butscher, "Scope of 1955 TC Arctic Activities in Greenland, Doc. No. 9–98–07–002, 1954" (Rigsarkivet, Udenrigsministeriet Collection, 105.F.9.A, US Aktivitet Udenfor Forsvarsområderne, 1952–1957); Steven J. Mock and Donald L. Alford, "CRREL Special Report 67: Installation of Ice Movement Poles in Greenland" (US Cold Regions Research and Engineering Laboratory, 1964). As many official documents about HIRAN are still classified, this section is based on open US military documents and reports from personnel, including original photos, diplomatic notes, and records kindly saved by former photomappers. Lack of access to restricted sources is a common complaint among historians interested in Cold War geodesy; for similar experiences, see Warner, "Political Geodesy: The Army, the Air Force and the World Geodetic System of 1960."

93. In addition to the photomappers, each HIRAN site team also included a medic and a survival specialist.

94. E.J. Mahoney, "1370th Personnel Taught to Live Anywhere from Pole to Equator," *Turner Air Force Base Reflex,* 1959; E.J. Mahoney, "1374th Fixes, Runs Electronic Equipment," *Turner Air Force Base Reflex,* 1959; Bill Sapp and Gordon Barnes, "1374th Mapping and Charting Squadron: A Tribute to Those Ground Station Guys," (http://www.1370th.org/1370pmw/1374/Tribute.htm). Additional details come from personal photos belonging to Pete Kessinger and C.J. Gregoire.

95. These details about the HIRAN camps on Greenland's ice sheet are courtesy of Roy P. Velvick, a staff sergeant who worked for the US 1370th Photomapping Group in Greenland and British Guiana (Roy P. Velvick, Personal Communication, 9 March 2013).

96. For floating ice islands, see William F. Althoff, *Drift Station: Arctic Outposts of Superpower Science* (Washington, DC: Potomac Books, 2007), 84ff.

97. Robert W. Gates, "Letter to Major Halken, Danish Liaison Officer, Sondrestrom Air Base, 16 March 1956" (Rigsarkivet, Udenrigsministeriet Collection, 105.F.9.A, US Aktivitet Udenfor Forsvarsområderne, 1952–1957). Capitalization in original.

98. "US Note No. 34, July 1952" (Rigsarkivet, Udenrigsministeriet Collection, UM 105.F.9); "Danish Note Verbale, 27 August 1952" (Rigsarkivet, Udenrigsministeriet Collection, UM 105.F.9).

99. Butscher, "Scope of 1955 TC Arctic Activities in Greenland, Doc. No. 9–98–07–002, 1954" (Rigsarkivet, Udenrigsministeriet Collection, 105.F.9.A, US Aktivitet Udenfor Forsvarsområderne, 1952–1957), 4; Torben Dithmer, "Notits: Planer for Amerikansk Virksomhed på Indlandsisen i 1955, P.J.V. J.Nr.105.F.9," 23 August 1954 (Rigsarkivet, Udenrigsministeriet Collection, 105.F.9.A, US Aktivitet Udenfor Forsvarsområderne, 1952–1957).

100. "Confidential Note, US Embassy in Copenhagen to the Ministry of Foreign Affairs Copenhagen, 25 April 1956" (Rigsarkivet, Udenrigsministeriet Collection, 105.F.9.A, US Aktivitet Udenfor Forsvarsområderne, 1952–1957); Vincent M. Miles, "Memorandum for Military Rights Branch, Department of the Air Force, AF00P_OP_S, Policy Division: Clearance for Scientific Projects in Greenland, 1955" (NARA, RG 341, Air Force Plans, Box 829, Folder 5); "Referat P.M.V. 15 August 1955" (Rigsarkivet, Udenrigsministeriet Collection, 105.F.9.A, US Aktivitet Udenfor Forsvarsområderne, 1952–1957); "Note Verbale, 18 June 1956, Ministry of Foreign Affairs Copenhagen to US Embassy in Copenhagen," Originally Confidential (Rigsarkivet, Udenrigsministeriet Collection, 105.F.9.A, US Aktivitet Udenfor Forsvarsområderne, 1952–1957).
101. Børge Fristrup, *The Greenland Ice Cap*, trans. David Stoner (Copenhagen: Rhodos, 1966), 23.
102. Cloud seeding works by introducing an agent (usually dry ice, silver iodide, or propane) which undergoes a natural phase change at a suitable temperature into a cloud in order to force the cloud droplets to form ice crystals and precipitate out as snow, thereby creating a clearing in an area which had previous been obscured. See Martin-Nielsen, "The Other Cold War: The United States and Greenland's Ice Sheet Environment, 1948–1966." These experiments in Greenland formed part of a much broader US Cold War weather modification program; see Kristine C. Harper, "Climate Control: United States Weather Modification in the Cold War and Beyond," *Endeavour* 32 (2008); Kristine C. Harper and Ronald E. Doel, "Environmental Diplomacy in the Cold War: Weather Control, the United States, and India, 1966–1967," in *Environmental Histories of the Cold War*, ed. John R. McNeill and Corinna R. Unger (Cambridge: Cambridge University Press, 2010); Ronald E. Doel and Kristine C. Harper, "Prometheus Unleashed: Science as a Diplomatic Weapon in the Lyndon B. Johnson Administration," *Osiris* 21 (2006); Charles C. Bates and John F. Fuller, *America's Weather Warriors, 1814–1985* (College Station, TX: Texas A&M University Press, 1986); James R. Fleming, *Fixing the Sky: The Checkered History of Weather and Climate Control* (New York: Columbia University Press, 2010).
103. 'Administrator' was the title given to SIPRE's chief civilian position. Gillis was also referred to as SIPRE's director.
104. James E. Gillis, "Letter to the Commander, Headquarters, Air Weather Service, MATS, 30 March 1956" (NSIDC, Carl S. Benson Collection, Greenland Snow Pit and Core Stratigraphy Subset, Box 2, HIRAN Folder).
105. R.G. Bounds, "Letter to the Director, SIPRE, 11 May 1956" (NSIDC, Carl S. Benson Collection, Greenland Snow Pit and Core Stratigraphy Subset, Box 2, HIRAN Folder).

106. HIRAN's Eismitte site was marked by a nine meter tall galvanized steel tower which had been installed by Project Jello in 1955.
107. Mock and Alford, "CRREL Special Report 67: Installation of Ice Movement Poles in Greenland," 2; "Northeast Air Command Headquarters, US Air Force to Paul-Emile Victor, 27 November 1956" (Relations Internationales, Etats-Unis: Suite Folder, Expéditions Polaires Francaises 1914–2001 Collection (20110210–257), Archives Nationales de France, Fontainebleau); Robert H. Lyddan, *Science in the Arctic Basin: A Report, Geodetic and Cartographic Aspects* (Washington, DC: US Committee on Polar Research, National Research Council and National Academy of Sciences, 1963), 35–36; Carl S. Benson and Richard H. Ragle, "SIPRE Special Report 19: Measurements by SIPRE on the Accumulation Markers of Expeditions Polaires Francaises in Central Greenland" (US Snow, Ice and Permafrost Research Establishment, 1956), 3–4.
108. Haefeli, "Projet de Programme Scientifique de l'EGIG, 3 September 1955" (Programmes Scientifiques EGIG I Folder, Expéditions Polaires Francaises 1914–2001 Collection (20110210–127), Archives Nationales de France, Fontainebleau). Also see Renaud, "La Participation de la Suisse à l'Expédition Glaciologique Internationale (EGIG) de 1957 à 1961."
109. That this mission was undertaken at a time when US military-scientific activities in Greenland were beginning to decline speaks to the high US interest in ice dynamics. For declining US activity on the island, see Martin-Nielsen, "'The Deepest and Most Rewarding Hole Ever Drilled': Ice Cores and the Cold War in Greenland"; Owen Wilkes and Jan Øberg, *Military Research and Development in Denmark and Greenland* (Lund, Sweden: Lund University Press, 1982).
110. Founded in 1961, the Cold Regions Research and Engineering Laboratory was SIPRE's successor organization.
111. Mock, "Greenland Operations of the 17th Tactical Airlift Squadron and CRREL."
112. Mock and Alford, "CRREL Special Report 67: Installation of Ice Movement Poles in Greenland," 5. Mock and Alford had difficulties locating the HIRAN stations because of snow accumulation which greatly exceeded SIPRE's predictions. Snow accumulation at the sites ranged from 30% to 80% more than anticipated—an error which underscores that understanding of Greenland's ice sheet remained tenuous.
113. Jean Nevière, "Nivellement Géodesique de l'Indlandsis, Campagne au Groënland 1948–1949–1950" (Rapports Scientifiques des Expéditions Polaires Francaises, 1954); George Wallerstein, "Movement Observations on the Greenland Ice Cap" (US Snow Ice and Permafrost Research Establishment, 1958); Benson and Ragle, "SIPRE Special Report 19: Measurements by SIPRE on the Accumulation Markers

of Expéditions Polaires Francaises in Central Greenland"; Louis Tschaen and Albert Bauer, "Le Mouvement de la Partie Centrale de l'Inlandsis du Groënland" (Association Internationale des Sciences Hydrologiques, 1958); Louis Tschaen, "Groënland 1948–1949–1950: Astronomie—Nivellement Géodesique sur l'Indlandsis, Nouveau Calcul" (Paris: Expéditions Polaires Francaises, 1959); Hermann Mälzer, "Die Höhenmessung bei der Sommerkampagne 1959 der Internationalen Glaziologischen Grönland-Expedition, 1960" (Programmes Scientifiques EGIG I Folder, Expéditions Polaires Francaises 1914–2001 Collection (20110210-127), Archives Nationales de France, Fontainebleau).

114. Børge Fristrup, "Recent Investigations of the Greenland Ice Cap," *Geografisk Tidsskrift* 58 (1959): 9.

115. The geodetic data obtained from the HIRAN stations was not openly available. However, the SHORAN and HIRAN systems themselves were not classified, and featured in books, academic journals, popular magazines, and scientific conferences beginning in 1946. SHORAN and HIRAN systems were also used by private companies including Virgil I. Kauffmann's Aero Service Corporation, which won surveying contracts from the US, Canadian, Spanish and Saudi Arabian governments, as well as private oil companies, through the 1950s and 1960s, and for whom Aslakson worked after his retirement from US service in 1955. For Aero Service, see the magazine *Prop Wash*, published out of the company's headquarters in Philadelphia. For an analysis of the relationship between US Cold War geodesy and secrecy, see Cloud, "American Cartographic Transformations During the Cold War."

116. For an overview, see Cornelis J. van der Veen et al., "Two Decades of Glaciological Investigations in South and Central Greenland," *Polar Geography* 24 (2000).

117. Albert Bauer, *Accélération de l'Ecoulement des Glaciers Groënlandais vers Leur Front et Détérmination de Leur Debit Solide* (Copenhagen: Meddelelser om Grønland, 1968), 75.

118. The 1967–1968 cooperative European expedition known as EGIG II straddled the boundary, using both conventional and modern geodetic techniques. This expedition had initially been planned as a full follow-up to the original EGIG expedition, but its scope was drastically restricted by the Danish authorities, who were angry with the original expedition's disregard for Danish interests (cf. chapter 4). See Joseph Vidal, *Mesures Effectuées par le Groupe de Géodesie B le Long du Profil Transversal Est-Ouest de l'Indlandsis du Groënland, EGIG, Eté 1967* (Copenhagen: Meddelelser om Grønland, 1983); Achim Karsten and Manfred Stober, "Deformation-Messungen auf dem Grönlandeschen Inlandeis Wahrend der Internationale Glaziologischen Grönland Expedition," *Polarforschung* 45 (1974).

119. Alice Jean Remington Drew, *Glacial Movements in Greenland from Doppler Satellite Observations* (Columbus, Ohio: Ohio State University

Institute of Polar Studies, 1983). For examples of other such studies, see Yushin Ahn and Jason E. Box, "Glacier Velocities from Time-Lapse Photos: Technique Development and First Results from the Extreme Ice Survey (EIS) in Greenland," *Journal of Glaciology* 56, no. 198 (2010); Ian Joughin et al., "Greenland Flow Variability from Ice-Sheet-Wide Velocity Mapping," *Journal of Glaciology* 56, no. 197 (2010).

120. As discussed in chapter 1, aircraft were not unknown in the polar regions at the time of Wegener's expedition, but Wegener chose not to employ them.

121. Neither Expéditions Polaires Françaises nor Expédition Glaciologique Internationale au Groënland had the expertise or equipment to safely land aircraft on the ice sheet, while for Project Jello, landed air support over such a long trek was considered too risky and unpredictable. For overviews of aviation in Greenland, see John Haile Cloe, *C-130D Support of the Greenland Ice Cap DYE Sites* (Elmerdorf Air Force Base, Alaska: Alaskan Air Command, 1977); William S. Carson, *Lifelines Through the Arctic* (New York: Duell, Sloan and Pearce, 1962); Edward Miller, "Ski Hercules," *Air Classics*, January 1975; "The Greenland Ice Plateau," *Air University Quarterly Review*, 1955. For a useful comparison of aviation in the Antarctic, see Henry M. Dater, *Aviation in the Antarctic* (Washington, DC: US Antarctic Office, 1963); George McCleary, "Air Transportation in the Antarctic," *US Antarctic Projects Office Bulletin* 5, no. 5 (1964).

122. Here, an *unprepared landing* refers to landing on a snow surface which had not been pre-prepared into a runway or landing strip. It is not intended to refer to the pilot's state of mind! The Soviets also developed the know-how to conduct unprepared air operations in the Arctic; see "The Soviet High Latitude Arctic Air Expedition and Drifting Stations in the Arctic Ocean," *Polar Record* 7 (1955); Terence Armstrong, *The Russians in the Arctic: Aspects of Soviet Exploration and Exploitation of the Far North, 1937–57* (Westport, CT: Greenwood Press, 1958).

123. For a vivid description of Hassell and Cramer's flight, see Gail S. Ravitts, "Bert R.J. 'Fish' Hassell and Parker D. 'Shorty' Cramer: Pilots of a Remarkable Rockford-to-Stockholm Flight," *Aviation History*, September 2000. Hassell went on to play a leading role in the construction of the US Sondrestrom and Thule Air Bases in Greenland.

124. "The Snowman Project, HQ Atlanta Division, ATC, Westover Field, Massachusetts, 1947" (NARA, RG 341, Air Force: Plans, Project Decimal File 1942–1954, Box 831, SG 581, TS), Section II.

125. Ibid., 5. Capitalization in original. For a summary of Operation Highjump, see Kenneth J. Bertrand, "A Look at Operation Highjump Twenty Years Later," *Antarctic Journal* 2 (1967): 5–12.

126. Cloe, *C-130D Support of the Greenland Ice Cap DYE Sites*, 13.

127. "The Snowman Project, HQ Atlanta Division, ATC, Westover Field, Massachusetts, 1947" (NARA, RG 341, Air Force: Plans, Project Decimal File 1942–1954, Box 831, SG 581, TS), 33.
128. For Beaudry's flight, see John L. Frisbee, "Valor: Greenland Rescue," *Air Force Magazine* 81, no. 3 (1998). Beaudry went on to author a key study on the air potential of Greenland's ice sheet for high latitude defense; see chapter 3. Further proof of concept came with the US military's 1949 Project Overheat and 1953 Project Mint Julep. For Project Overheat, see "CRREL Report No. ACFEL TR 27: Project Overheat Final Report" (US Cold Regions Research and Engineering Laboratory, 1950). For Project Mint Julep, see "ADTIC Publication A-104a: Project Mint Julep—Investigation of Smooth Areas of the Greenland Ice Cap, 1953, Part 1" (Research Studies Institute, Air University, Maxwell Air Force Base, Alabama: Arctic, Desert, Tropic Information Center, 1955).
129. Edison Blair, *Arctic Adventure: An Account of the First C-47 Landing at the North Pole* (US Air Force Monograph, 1952).
130. Known as N-33 and N-34, the radar stations were designed to extend Thule's radar coverage and were serviced by the 931st Aircraft Control and Warning Squadron.
131. Mock and Alford, "CRREL Special Report 67: Installation of Ice Movement Poles in Greenland," A-1.
132. Key examples of C-130 support on Greenland's ice sheet include the 1966–1967 Danish-American Blue Ice seismological project at Inge Lehmann Station in north Greenland, a revisit to the collapsed Camp Century in 1969, and the joint Danish-Swiss-American Greenland Ice Sheet Project (GISP) ice core work of the 1970s.
133. Dater, *Aviation in the Antarctic*, U1.
134. Bill Duncan, "New Breed of Pilot Tames the Arctic" (*Long Beach Press-Telegram*, p. B-2, 1964).
135. The planes were also a maintenance nightmare: with the addition of skis, the time to change a tire on the airplanes increased from 45 minutes to ten hours.
136. Fritz Loewe, "The Scientific Exploration of Greenland from the Norsemen to the Present, Paper presented at a Public Lecture on May 26, 1969 at Ohio State University" (Columbus, OH: Institute of Polar Studies, Ohio State University, 1970), 12–13.
137. The most complete resource on the DEW Line is the impressive P. Whitney Lackenbauer, Matthew Farish, and Jennifer Arnold-Lackenbauer, eds., *The Distant Early Warning (DEW) Line: A Bibliography and Documentary Resource List* (Calgary: Arctic Institute of North America, 2005).
138. Greenland also hosted two coastal DEW Line stations, called DYE-1 and DYE-4. Greenland's four DEW Line stations were active until 1988–1991.
139. Cloe, *C-130D Support of the Greenland Ice Cap DYE Sites*, 17.

140. For Greenland's DYE stations, see Harold B. Goyette, "DEW Line Canada-Iceland Link," *The Military Engineer* 361, Sept-Oct (1962); Cloe, *C-130D Support of the Greenland Ice Cap DYE Sites.*
141. "Meeting in Copenhagen, 8–9 May 1961" (Rigsarkivet, 0030, Grønlandsministeriet Collection, Journalsager 1957–1989, Box 2083, Folder Paul-Emile Victor, Ekspeditioner, 1420-21-01).
142. See, for example, Stephen E. Ragone and C.A. Wolf, "CRREL Report no. SR 169: Analysis of the Major Cationic Constituents of the 1964 to 1969 Snow Accumulations at DYE Sites 2 and 3, Greenland" (US Cold Regions Research and Engineering Laboratory, 1972); James H. Cragin, M.M. Herron, and Chester C. Langway, "CRREL Report no. RR 341: Chemistry of 700 Years of Precipitation at DYE 3, Greenland" (US Cold Regions Research and Engineering Laboratory, 1975); Steven J. Mock, "Geodetic Positions of Borehole Sites of the Greenland Ice Sheet Project" (US Cold Regions Research and Engineering Laboratory, 1976).
143. Dansgaard, *Frozen Annals: Greenland Ice Cap Research*, 66.
144. Chester C. Langway, "ERDC/CRREL Report TR-08-1: The History of Early Polar Ice Cores" (US Cold Regions Research and Engineering Laboratory, 2008), 28.
145. The deepest ice core in the world at the time of GISP was the 2,164 meter long bedrock core drilled at Byrd Station, Antarctica, in 1966–1968.
146. Loewe, "The Scientific Exploration of Greenland from the Norsemen to the Present, Paper presented at a Public Lecture on May 26, 1969 at Ohio State University," 12–13. Loewe added in a qualification that Station Eismitte was the coldest scientific station on earth during the 1930–1931 German overwinter; it has since been surpassed in this aspect by scientific stations in the Antarctic.

Epilogue   A Conspicuous Absence

1. Greenland as a "part of" Denmark requires some clarification. Until 1953, Greenland was a Danish colony. In that year, the island was incorporated as a Danish county (*amt*) and a period of so-called Danization began. In 1979, with Home Rule, Greenland gained limited autonomy, and in 2009 the island assumed self-determination (that is, self-government for policing, judicial affairs, and natural resources). See Mark Nuttall, "Self-Rule in Greenland: Towards the World's First Independent Inuit State?," *Indigenous Affairs* 3–4 (2008); Erik Beukel and Frede P. Jensen, *Phasing Out the Colonial Status of Greenland* (Copenhagen: Meddelelser om Grønland, 2010).
2. A second notable absence is the lack of Greenlanders (or *Kalaallit*) after Rasmus Villumsen's death in 1930. This absence stems neither from deliberate omission nor neglect; rather, apart from Villumsen,

Greenlanders were not present at Eismitte, and played only marginal roles in the stories told in this book. In northwestern Greenland, for example, communities of Greenlanders were forcibly relocated to make way for the US Thule Air Base, which then acted as a staging point for the Jello expedition. While the role of indigenous people in the history of science is topical, important, and understudied, in the case at hand, Greenlanders are too tenuously connected to Eismitte to form an integral part of my narrative. Their story in relation to scientific investigation in Greenland is slowly beginning to be told, but much work remains. See, for example, Henry Nielsen and Henrik Knudsen, "Cold Atoms: The Hunt for Uranium in Greenland in the Late Cold War, 1970–1989" (forthcoming).
3. Here, mainland Norway refers to the western Scandinavian Peninsula. It excludes Norway's overseas possessions.
4. Christopher Jacob Ries, "Armchairs, Dogsleds, Ships and Airplanes: Field Access, Scientific Credibility, and Geological Mapping in Northern and North-Eastern Greenland, 1900–1939," in *Scientists and Scholars in the Field: Studies in the History of Fieldwork and Expeditions*, ed. Kristian Hvidtfelt Nielsen and Christopher Jacob Ries (Aarhus: Aarhus University Press, 2012), 329.
5. The intervening years were littered with diplomatic negotiations and commercial (namely, whaling, fishing, and sealing) disputes between the two countries; see Einar-Arne Drivenes and Harald Dag Jølle, "The History of Norway in the Polar Regions," in *Into the Ice: The History of Norway and the Polar Regions*, ed. Einar-Arne Drivenes and Harald Dag Jølle (Oslo: Gyldendal Norsk Forlag, 2006), 296ff.
6. "Statut Juridique du Groënland Oriental, Cour Permanante de Justice Internationale, 26ième Session, 5 avril 1933" (Leydon: A.W. Sijthoff's Publishing Co., 1933).
7. Ries, "Armchairs, Dogsleds, Ships and Airplanes: Field Access, Scientific Credibility, and Geological Mapping in Northern and North-Eastern Greenland, 1900–1939," 329.
8. Ibid.
9. For relevant discussions, see P. Whitney Lackenbauer, ed. *Canada and Arctic Sovereignty and Security: Historical Perspectives* (Calgary: University of Calgary Press—Calgary Papers in Military and Strategic Studies, 2011); Peder Roberts, "Science and Political Authority in the Polar Regions in the Cold War—and Beyond" (Talk given at Aarhus University, February 25, 2013); Matthias Heymann and Ronald E. Doel, eds., *Exploring Greenland: Science and Technology in Cold War Settings* (forthcoming). For environmental authority, see Adrian Howkins, "A Formal End to Informal Imperialism: Environmental Nationalism, Sovereignty Disputes, and the Decline of British Interests in Argentina, 1933–1955," *British Scholar Journal* 2 (2010).

10. These are key ideas in northern diplomatic history more broadly; for the Canadian case, for example, see Stephen Bocking, "A Disciplined Geography: Aviation, Science and the Cold War in Northern Canada," *Technology and Culture* 50 (2009).
11. Peder Roberts, "Nordic or National? Postwar Visions of Polar Conflict and Cooperation," in *Science, Geopolitics and Culture in the Polar Regions: Norden Beyond Borders*, ed. Sverker Sörlin (Surrey, England: Ashgate, 2013), 55.
12. Nikolaj Petersen, "The Politics of US Military Research in Greenland in the Early Cold War," *Centaurus* 55 (2013): 294–318; Niels Amstrup, "Grønland i det Amerikansk-Danske Forhold, 1945–1948," in *Studier i Dansk Udenrigspolitik: Tilegnet Erling Bjøl*, ed. Niels Amstrup and Ib Faurby (Aarhus: Politica, 1978), 175. For Canada's similar postwar experience with US Arctic weather stations, see Daniel Heidt, "Clenched in the JAWS of America? Canadian Sovereignty and the Joint Arctic Weather Stations, 1946–1972," *Calgary Papers in Military and Strategic Studies* (Canadian Arctic Sovereignty and Security: Historical Perspectives) No. 4, 2011.
13. Amstrup, "Grønland i det Amerikansk-Danske Forhold, 1945–1948," 167.
14. Roberts, "Nordic or National? Postwar Visions of Polar Conflict and Cooperation," 9.
15. "Defense of Greenland 1951, Agreement Between the United States and the Kingdom of Denmark, April 27, 1951" (American Foreign Policy 1950–1955, Basic Documents, Department of State Publication 6446, General Foreign Policy Series 117, Washington DC).
16. Eske Brun, "Greenland," *Arctic* 19 (1966): 63.
17. "Nordic or National? Postwar Visions of Polar Conflict and Cooperation," 9. Emphasis added.
18. Paul-Emile Victor, "Letter to Monsieur De Guebriant, Ambassade de France au Danemark, 17 February 1949, Ref. O. 1821" (Relations avec le Danemark 1947–1991 Folder, Expéditions Polaires Francaises 1914–2001 Collection (20110210–255), Archives Nationales de France, Fontainebleau); "Une Maison Construite au Milieu de l'Indlandsis et Offerte au Danemark: l'Explorateur Francais Paul-Emile Victor Offre au Groënland une Station Scientifique," *Berlingske Aftenavis* (copy), 4 January 1950 (Relations avec le Danemark 1947–1991 Folder, Expéditions Polaires Francaises 1914–2001 Collection (20110210–255), Archives Nationales de France, Fontainebleau).
19. It is hard (and, indeed, fruitless) to neatly separate glaciology, geology, and cartography on these expeditions; near Greenland's coasts, the three go hand-in-hand.
20. Børge Fristrup, "Recent Investigations of the Greenland Ice Cap," *Geografisk Tidsskrift* 58 (1959): 11; Niels Nielsen, "Danish National Committee for the International Geophysical Year: Plan

for the Glaciological Researches in Greenland 1956–1958" (NARA, RG 27, Entry 3, Geophysical Year 1953–1960, 130/16/9/2, Box 3: Glaciology); Børge Fristrup, "Studies of Four Glaciers in Greenland," *Geografisk Tidsskrift* 59 (1960).
21. While Denmark, as the host country, had priority for filling the expedition's scientific posts, in practice the historical Danish glaciological focus on the coastal regions meant that Denmark did not have the expertise needed for ice sheet work (cf. chapter 4).
22. "Meeting in Copenhagen, 8–9 May 1961" (Rigsarkivet, 0030, Grønlandsministeriet Collection, Journalsager 1957–1989, Box 2083, Folder Paul-Emile Victor, Ekspeditioner, 1420–21–01).
23. See, for example, the assessment of F.J.G. Cunningham, director of the Administration and Lands Branch of Canada's Department of Northern Affairs and Natural Resources, in a memo to his minister (F.J.G. Cunningham, "Memorandum for the Minister: Mr. Eske Brun, Northern Administration and Lands Branch, Ottawa, 30 July 1957" (LAC, RG85, Denmark and Greenland General File No. 1005–7, Vol. 10, Department of Northern Affairs and National Resources, ATIP Division Interim Box 292–2001099652), 3). For Brun's approach to Greenland, see Jens Heinrich, *Eske Brun og det Moderne Grønlands Tilblivelse 1932–1964* (PhD Thesis, Ilisimatusarfik/Grønlands Universitet, 2010).
24. "Meeting in Copenhagen, 8–9 May 1961" (Rigsarkivet, 0030, Grønlandsministeriet Collection, Journalsager 1957–1989, Box 2083, Folder Paul-Emile Victor, Ekspeditioner, 1420–21–01).
25. Børge Fristrup, "Noter til Mødet Med EGIG D. 8–9/5–1961" (Rigsarkivet, 0030, Grønlandsministeriet Collection, Journalsager 1957–1989, Box 2083, Folder Paul-Emile Victor, Ekspeditioner, 1420–21–01).
26. "Meeting in Copenhagen, 8–9 May 1961" (Rigsarkivet, 0030, Grønlandsministeriet Collection, Journalsager 1957–1989, Box 2083, Folder Paul-Emile Victor, Ekspeditioner, 1420–21–01). Nielsen followed up his comments with a strongly worded letter to Albert Bauer rebuking the Frenchman for his continued efforts to organize scientific work in Greenland (Niels Nielsen, "Letter to Albert Bauer, 6 July 1961" (Rigsarkivet, 0030, Grønlandsministeriet Collection, Journalsager 1957–1989, Box 2083, Folder Paul-Emile Victor, Ekspeditioner, 1420–21–01)).
27. "Meeting in Copenhagen, 8–9 May 1961" (Rigsarkivet, 0030, Grønlandsministeriet Collection, Journalsager 1957–1989, Box 2083, Folder Paul-Emile Victor, Ekspeditioner, 1420–21–01).
28. See, for example Edward Muller, "The Top of the World," *The Reader's Digest*, March 1954; Richard J. Russell, "Instability of Sea Ice," *American Scientist* 45, no. 5 (December, 1957).
29. Eske Brun, "Letter to Albert Bauer, 22 March 1961" (Rigsarkivet, 0030, Grønlandsministeriet Collection, Journalsager 1957–1989,

Box 2083, Folder Børge Fristrup, 1420-21-02). Also see Fristrup, "Noter til Mødet Med EGIG D. 8-9/5-1961" (Rigsarkivet, 0030, Grønlandsministeriet Collection, Journalsager 1957-1989, Box 2083, Folder Paul-Emile Victor, Ekspeditioner, 1420-21-01).

30. The others were Hans Oeschger at Bern University and Chester C. Langway at the US Army's Snow, Ice and Permafrost Research Establishment.
31. Herbert F. Feaver, "Greenland Diary, Wednesday, July 10, 1957" (LAC, RG85, Denmark and Greenland General File No. 1005-7, Vol. 10, Department of Northern Affairs and National Resources, ATIP Division Interim Box 292-2001099652).
32. "Stricken Dane Evacuated" (Press Release from Sondrestrom Air Base, Greenland, 1 February 1963).
33. See, for example, "Dyess Crew Aids in Mercy Flight" (Newspaper article courtesy of Don Wilkerson, 17th Troop Carrier Squadron Pilot, May 1964); "Memo to the Deputy Minister, Department of Northern Affairs and National Resources, Ottawa, from the Under Secretary of State for External Affairs, 7 April 1955" (LAC, RG85, Denmark and Greenland General File No. 1005-7, Vol. 7, Department of Northern Affairs and National Resources, ATIP Division Interim Box 46-2001099652).
34. Fristrup, "Noter til Mødet Med EGIG D. 8-9/5-1961" (Rigsarkivet, 0030, Grønlandsministeriet Collection, Journalsager 1957-1989, Box 2083, Folder Paul-Emile Victor, Ekspeditioner, 1420-21-01). Here, Fristrup is referring specifically to snow marker measurements.
35. For the Camp Century deep ice core, see Willi Dansgaard and et al., "One Thousand Centuries of Climatic Record from Camp Century on the Greenland Ice Sheet," *Science* 166 (1969). For the glacial climatic leaps (or Dansgaard-Oeschger events), see Chester C. Langway, Hans Oeschger, and Willi Dansgaard, eds., *Greenland Ice Cores: Geophysics, Geochemistry and Environment* (Washington, DC: American Geophysical Union, 1985). And for the Rolls Royce drill, see Willi Dansgaard, *Frozen Annals: Greenland Ice Cap Research* (Copenhagen: Niels Bohr Institute, 2005), 76; Sigfus J. Johnsen, "A Fast Lightweight Core Drill," *Journal of Glaciology* 25 (1980). This group of young, technically-minded scientists included Niels Gundestrup, Claus Hammer, Steffen Hansen, Niels Reeh and Icelander Sigfus Johnsen. For Dansgaard's laboratory, see Michele Citterio, "In Memoriam: Prof. Dr. Willi Dansgaard, 1922-2011," *Studia UBB Geologia* 56 (2011).
36. Denmark was involved with, among others, the 1966 Camp Century project (a joint Denmark-US undertaking), the 1971-1981 Greenland Ice Sheet Project (Denmark, Switzerland and the United States), the 1989-1992 Greenland Ice Core Project (Denmark, Germany, Iceland, France, Belgium, Italy, Switzerland and the United Kingdom), and the 1996-2004 North Greenland Ice Core Project (Denmark, Germany,

Japan, Sweden, Switzerland, France, Belgium, Iceland, and the United States).
37. For discussion, see Maiken Lolck, *Klima, Kold Krig og Iskerner* (Aarhus: Aarhus Universitetforlag, 2006).
38. For the title 'King of Greenland', see Marianne Krogh Andersen, "Kongen af Grønland," *Weekendavisen* (February 4, 2011), 8–9.

Bibliography

"ADTIC Publication A-104a: Project Mint Julep – Investigation of Smooth Areas of the Greenland Ice Cap, 1953, Part 1." Research Studies Institute, Air University, Maxwell Air Force Base, Alabama: Arctic, Desert, Tropic Information Center, 1955.

Ahlmann, Hans W. "Review of Scientific Results of the German Alfred Wegener Greenland Expedition 1929 and 1930–31." *Geografiska Annaler* 23 (1941): 134–37.

Ahlmann, Hans W. "The Contribution of Polar Expeditions to the Science of Glaciology (A Lecture Delivered to the Scott Polar Research Institute on 1 May 1948)." *Polar Record* 5 (1949): 324–31.

Ahn, Yushin, and Jason E. Box. "Glacier Velocities from Time-Lapse Photos: Technique Development and First Results from the Extreme Ice Survey (EIS) in Greenland." *Journal of Glaciology* 56, no. 198 (2010): 723–34.

Althoff, William F. *Drift Station: Arctic Outposts of Superpower Science.* Washington, DC: Potomac Books, 2007.

Amdrup, Georg C. *Danmark-Ekspeditionen til Grønlands Nordøstkyst 1906–1908 Under Ledelse af L. Mylius-Erichsen.* Copenhagen: Meddelelser om Grønland, 1913.

Amstrup, Niels. "Grønland i det Amerikansk-Danske Forhold, 1945–1948." In *Studier i Dansk Udenrigspolitik: Tilegnet Erling Bjøl*, edited by Niels Amstrup and Ib Faurby, 155–98. Aarhus: Politica, 1978.

Andersen, Marianne Krogh. "Kongen af Grønland." *Weekendavisen*, February 4, 2011, pp. 8–9.

"APCS: An Exceptional Unit." *The MATS Flyer* 11 (1964): 10–12.

"Appendix 1: Importance of the High Arctic to North American Defense." In *Report of the Arctic Institute of North America, Presented at the Hearings before the Committee on Merchant Marine and Fisheries, House of Representatives 85th Congress, 2nd Session, 22–24 January 1958.* New York: Office of Naval Research, Arctic Research Advisory Committee, October 25, 1957.

Archer, Clive. "The United States Defense Areas in Greenland." *Cooperation and Conflict* 23 (1988): 123–44.

Armstrong, Terence. *The Russians in the Arctic: Aspects of Soviet Exploration and Exploitation of the Far North, 1937–57.* Westport, CT: Greenwood Press, 1958.

Aslakson, Carl I. *The Principles of SHORAN Mapping*. Philadelphia, PA: Aero Service Corporation, Photogrammetic Engineers, 1957.
Aslakson, Carl I., and Donald A. Rice. "The Use of SHORAN in Geodetic Control." *American Geophysical Union Transactions* 27 (1946): 459–63.
Bader, Henri. "SIPRE Research Report 2, AD-014 366: Sorge's Law of Densification of Snow on High Polar Glaciers." US Snow, Ice and Permafrost Research Establishment, 1953.
Bader, Henri. "Sorge's Law of Densification of Snow on High Polar Glaciers." *Journal of Glaciology* 2, no. 15 (1954): 319–23.
Bader, Henri. "United States Polar Ice and Snow Studies in the International Geophysical Year." In *Geophysics and the IGY*, edited by Hugh Odishaw and Stanley Ruttenberg, 177–81. Washington, DC: American Geophysical Union, 1958.
Baigent, Elizabeth. "'Deeds Not Words'? Life Writing and Early 20th Century British Polar Exploration." In *New Spaces of Exploration: Geographies of Discovery in the 20th Century*, edited by Simon Naylor and James R. Ryan, 23–51. London: I.B. Tauris, 2010.
Balchen, Bernt, Corey Ford, and Oliver LaFarge. *War Below Zero: The Battle for Greenland*. New York: Houghton Mifflin Co., 1944.
Baldwin, Hanson W. "Arctic Repels Warfare: US Exercises Show Great Difficulties, Make Large-Scale Operations Unlikely." *The New York Times*, February 8, 1948.
Barillon, E. "Correspondance." *Journal of Glaciology* 1, no. 6 (1949): 337.
Bates, Charles C., and John F. Fuller. *America's Weather Warriors, 1814–1985*. College Station, TX: Texas A&M University Press, 1986.
Bauer, Albert. *Accélération de l'Ecoulement des Glaciers Groënlandais Vers Leur Front et Détérmination de Leur Debit Solide*. Copenhagen: Meddelelser om Grønland, 1968.
Becks, Peter J. *The International Politics of Antarctica*. London: Routledge, 1986.
Belanger, Dian O. *Deep Freeze: The United States, the International Geophysical Year, and the Origins of Antarctica's Age of Science*. Boulder, CO: University of Colorado Press, 2006.
Benson, Carl S. "CRREL MP 664: Stratigraphic Studies in the Snow and Firn of the Greenland Ice Sheet." US Cold Regions Research and Engineering Laboratory, 1961.
Benson, Carl S. "CRREL Research Report 70: Stratigraphic Studies in the Snow and Firn of the Greenland Ice Sheet." US Cold Regions Research and Engineering Laboratory, 1962.
Benson, Carl S. "Oral History Interview with Karen Brewster." Byrd Polar Research Center Oral History Program, Ohio State University, June 22, 2001.
Benson, Carl S. "SIPRE Report 24: Scientific Work of Party Crystal, 1954 (Preliminary Report, April 1955)." In *Corps of Engineers, Greenland Ice Cap Research Program, Studies Completed in 1954*. Vicksburg, MS: Army-MRC, 1955.

Benson, Carl S. "SIPRE Report 25: Operations and Logistics of Ice-Cap Party Crystal (April 1955)." In *Corps of Engineers, Greenland Ice Cap Research Program, Studies Completed in 1954*. Vicksburg, MS: Army-MRC, 1955.

Benson, Carl S. "SIPRE Special Report 17: Resupply of Ice-Cap Expeditions by Airdrop." US Snow, Ice and Permafrost Research Establishment, 1955.

Benson, Carl S. "Stratigraphic Studies in the Snow and Firn of the Greenland Ice Sheet." California Institute of Technology, 1960.

Benson, Carl S., and Richard H. Ragle. "SIPRE Special Report 18: Project Jello, SIPRE Greenland Expedition 1955, Report on Special Foods Provided by the Quartermaster Food and Container Institute." US Snow, Ice and Permafrost Research Establishment, 1956.

Benson, Carl S., and Richard H. Ragle. "SIPRE Special Report 19: Measurements by SIPRE on the Accumulation Markers of Expéditions Polaires Françaises in Central Greenland." US Snow, Ice and Permafrost Research Establishment, 1956.

Benson, Carl S., and Charles R. Wilson. "Barry Bishop's Research on the Sheer Moraines in the Thule Area, Northwest Greenland." *Mountain Research and Development* 16, no. 3 (1996): 309–11.

Bertrand, Kenneth J. "A Look at Operation Highjump Twenty Years Later." *Antarctic Journal* 2 (1967): 5–12.

Beukel, Erik, and Frede P. Jensen. *Phasing out the Colonial Status of Greenland*. Copenhagen: Meddelelser om Grønland, 2010.

Blair, Edison. *Arctic Adventure: An Account of the First C-47 Landing at the North Pole*. US Air Force Monograph, 1952.

Bobe, Louis. *Diplomatarium Groenlandicum 1492–1814*. Copenhagen: Meddelelser om Grønland, 1936.

Bocking, Stephen. "A Disciplined Geography: Aviation, Science and the Cold War in Northern Canada." *Technology and Culture* 50 (2009): 265–90.

Bouché, Michel. *Groënland: Station Centrale*. Paris: Bernard Grasset, 1952.

Bright, J. Christopher. *Continental Defense in the Eisenhower Era: Nuclear Antiaircraft Arms and the Cold War*. New York: Palgrave Macmillan, 2010.

Brockamp, Bernhard, Ernst Sorge, and Kurt Wölcken. *Wissenschaftliche Ergebnisse der Deutschen Grönland-Expedition Alfred Wegener 1929 und 1930/1931, Band II: Seismik*. Leipzig: F.A. Brockhaus, 1933.

Brown, Robert. "Obituary: Dr Hendrik Rink [sic]." *The Geographical Journal* 3 (1894): 65–67.

Brun, Eske. "Greenland." *Arctic* 19 (1966): 62–69.

Brun, Eske. *Mit Grønlandsliv: Erindringer af Eske Brun*. Denmark: Gyldendal, 1985.

Brøsted, Jens, and Mads Fægteborg. *Thule: Fangerfolk og Militæranlæg*. Copenhagen: Akademisk Forlag, 1985.

Burkhardt, Richard. *Patterns of Behavior: Konrad Lorenz, Niko Tinbergen, and the Founding of Ethology*. Chicago, IL: University of Chicago Press, 2005.

"Byrd Stresses Use of Arctic in a War." *The New York Times*, November 18, 1947.

Cailleux, André. "Premiers Enseignements Glaciologiques des Expéditions Polaires Francaises, 1948–1951." Expéditions Polaires Francaises, 1951.

"Captain Koch's Crossing of Greenland." *Bulletin of the American Geographical Society* 46 (1914): 356–60.

Carey, Mark. "The History of Ice: How Glaciers Became an Endangered Species." *Environmental History* 12 (2007): 497–527.

Carson, William S. *Lifelines Through the Arctic.* New York: Duell, Sloan and Pearce, 1962.

Caswell, John Edwards. *Arctic Frontiers: United States Explorations in the Far North.* Norman, Oklahoma: University of Oklahoma Press, 1956.

Chaturvedi, Sanjay. *The Polar Regions: A Political Geography.* Chichester, UK: John Wiley & Sons, 1996.

Christie, Robert W. "Bacterial Variations in the Nasopharynx and Skin of Isolated Arctic Scientists." *New England Journal of Medicine* 258 (1958): 531–33.

Christie, Robert W. "Medical Notes on a Greenland Ice Cap Expedition." *Journal of the American Medical Association* 164 (1957): 1314–17.

Christie, Robert W. "Pioneer in Pathology." *Dartmouth Medicine* 27 (2003): 50–57.

Citterio, Michele. "In Memoriam: Prof. Dr. Willi Dansgaard, 1922–2011." *Studia UBB Geologia* 56 (2011): 43–44.

Cloe, John Haile. *C-130D Support of the Greenland Ice Cap DYE Sites.* Elmerdorf Air Force Base, AK: Alaskan Air Command, 1977.

Cloud, John. "American Cartographic Transformations During the Cold War." *Cartography and Geographic Information Science* 29, no. 3 (2002): 261–82.

Cloud, John. "Hidden in Plain Sight: The CORONA Reconnaissance Satellite Program and the Cold War Convergence of the Earth Sciences." *Annals of Science* 58 (2001): 203–09.

Coates, Ken, P. Whitney Lackenbauer, William R. Morrison, and Greg Poelzer, eds. *Arctic Front: Defending Canada in the Far North.* Toronto: Thomas Allen & Son, 2008.

Coates, Ken, and William R. Morrison. *The Alaska Highway in World War II: The US Army of Occupation in Canada's Northwest.* Toronto: University of Toronto Press, 1992.

Coletta, Paolo, ed. *United States Navy and Marine Corps Bases Overseas.* Westport, CT: Greenwood Press, 1985.

Collis, Christy, and Klaus Dodds. "Assault on the Unknown: The Historical and Political Geographies of the International Geophysical Year, 1957–1958." *Journal of Historical Geography* 34 (2008): 555–73.

"Conference on Glaciological Research Under the Auspices of the Arctic Institute of North America and the American Geographical Society, New York City." American Geographical Society Report, 1949.

Cornett, Lloyd H., and Mildred W. Johnson. *A Handbook of Aerospace Defense Organization 1946–1980.* Colorado: Office of History, Aerospace Defense Center, Peterson Air Force Base, 1980.

BIBLIOGRAPHY

Cragin, James H., M.M. Herron, and Chester C. Langway. "CRREL Report No. RR 341: Chemistry of 700 Years of Precipitation at DYE 3, Greenland." US Cold Regions Research and Engineering Laboratory, 1975.

"CRREL Report No. ACFEL TR 27: Project Overheat Final Report." US Cold Regions Research and Engineering Laboratory, 1950.

Dansgaard, Willi. *Frozen Annals: Greenland Ice Cap Research*. Copenhagen: Niels Bohr Institute, 2005.

Dansgaard, Willi, Sigfus J. Johnsen, Jesper Møller, and Chester C. Langway. "One Thousand Centuries of Climatic Record from Camp Century on the Greenland Ice Sheet." *Science* 166 (1969): 377–81.

Dansgaard, Willi, and Anker Weidick. "Klimaforværring i Grønland?" *Tidsskriftet Grønland* (November 1965): 399–405.

Darwent, John, Christyann Darwent, Genevieve LeMoine, and Hans Lange. "Archaeological Survey of Eastern Inglefield Land, Northwest Greenland." *Arctic Anthropology* 44 (2007): 51–86.

Dater, Henry M. *Aviation in the Antarctic*. Washington, DC: US Antarctic Office, 1963.

Daugherty, Charles Michael. *City Under the Ice: The Story of Camp Century*. New York: The Macmillan Company, 1963.

Davies, John. "Uncle Joe Knew Where You Lived, Part 1." *Sheetlines: Journal of the Charles Close Society for the Study of Ordnance Survey Maps* 72, April (2005): 26–38.

Davies, John. "Uncle Joe Knew Where You Lived, Part 2." *Sheetlines: Journal of the Charles Close Society for the Study of Ordnance Survey Maps* 73, August (2005): 1–15.

de Bont, Raf. "Between the Laboratory and the Deep Blue Sea: Space Issues in the Marine Stations of Naples and Wimereux." *Social Studies of Science* 39 (2009): 199–227.

de Caunes, Georges. "Comment Nous Avons Etabli un Pont 'Aérien' Entre l'Islande et le Groënland." *Paris-Presse*, August 16, 1949.

de Caunes, Georges. *Imarra: Aventures Groënlandaises*. Paris: Editions Hoebeke, 1998.

de Caunes, Georges. "Je Reviens du Grand Nord." *Radiodiffusion Francaise*, 1949.

De Lacour, Geneviève. "Retour au Groenland 70 Ans Après Paul-Emile Victor." *National Geographic French Edition* (March 2007): 4–23.

de Quervain, Marcel. *Schneekundliche Arbeiten der Internationalen Glaziologischen Grönlandexpedition: Nivologie*. Copenhagen: Meddelelser om Grønland, 1969.

"Defense Mapping Agency Technical Report 80–003: Geodesy for the Layman." Washington, DC: Defense Technical Information Center, 1984.

Demhardt, Imre J. "Alfred Wegener's Hypothesis on Continental Drift and Its Discussion in Petermanns Geographische Mitteilungen, 1912–1942." *Polarforschung* 75 (2005): 29–35.

Diamond, Jared. *Collapse: How Societies Choose to Fail or Succeed*. New York: Penguin Books, 2005.

Dibb, Jack E., and Henrik B. Clausen. "A 200-Year Pb-210 Record from Greenland." *Journal of Geophysical Research* 102, no. D4 (1997): 4325–32.
Dille, John. "Mapping the Earth." *Life Magazine* (1965): 124–38.
Doel, Ronald E. "Constituting the Postwar Earth Sciences: The Military's Influence on the Environmental Sciences in the USA after 1945." *Social Studies of Science* 33 (2003): 635–66.
Doel, Ronald E. "Quelle Place Pour les Sciences de l'Environnement Physique dans l'Histoire Environnementale?" *Revue d'Histoire Moderne et Contemporaine* 56 (2009): 137–64.
Doel, Ronald E., and Kristine C. Harper. "Prometheus Unleashed: Science as a Diplomatic Weapon in the Lyndon B. Johnson Administration." *Osiris* 21 (2006): 66–85.
Drew, Alice Jean Remington. *Glacial Movements in Greenland from Doppler Satellite Observations*. Columbus, OH: Ohio State University Institute of Polar Studies, 1983.
Drivenes, Einar-Arne, and Harald Dag Jølle. "The History of Norway in the Polar Regions." In *Into the Ice: The History of Norway and the Polar Regions*, edited by Einar-Arne Drivenes and Harald Dag Jølle. Oslo: Gyldendal Norsk Forlag, 2006.
Driver, Felix. "Modern Explorers." In *New Spaces of Exploration: Geographies of Discovery in the 20th Century*, edited by Simon Naylor and James R. Ryan, 241–49. London: I.B. Tauris, 2010.
Dubois, Georges. "Données Numériques Relatives aux Glaciations Quaternaires." *Bulletin de l'Association de l'Institut des Sciences Géologiques de Strasbourg* (1931): 30–36.
Dumont d'Urville, Jules Sébastian César. *Voyage au Pôle Sud et dans l'Océanie sur les Corvettes l'Astrolabe et la Zélée, 1837–1840*. Paris: Gide et J. Baudry, 1854.
Duncan, Bill. "New Breed of Pilot Tames the Arctic." (*Long Beach Press-Telegram*, p. B-2, 1964).
"DUPI Vol. 1, Grønland Under den Kolde Krig: Dansk og Amerikansk Sikkerhedspolitik, 1945–68." Copenhagen: Dansk Udenrigspolitisk Institut, 1997.
Dupree, A. Hunter. *Science in the Federal Government: A History of Policies and Activities to 1940*. Cambridge, MA: Harvard University Press, 1957.
"Dyess Crew Aids in Mercy Flight." Newspaper article courtesy of Don Wilkerson, 17th Troop Carrier Squadron Pilot, May 1964.
Einhorn, Eric S. *National Security and Domestic Politics in Post-War Denmark: Some Principal Issues, 1945–1961*. Odense, Denmark: Odense University Press, 1975.
Einhorn, Eric S. "The Reluctant Ally: Danish Security Policy 1945–49." *Journal of Contemporary History* 10, no. 3 (1975): 493–512.
Ellingsen, Ellman. "The Military Balance on the Northern Flank." In *Clash in the North: Polar Summitry and NATO's Northern Flank*, edited by Walter Goldstein. Washington, DC: Pergamon-Brassey's, 1988.

Elzinga, Aant. "Some Aspects in the History of Ice Core Drilling and Science from IGY to EPICA." 3rd SCAR Antarctic History Action Group Workshop, Byrd Polar Research Center, Ohio State University, October 2007.

Emmanuel, Marthe. *La France et l'Exploration Polaire: De Verrazano à la Perouse, 1583–1788.* Paris: Nouvelles Editions Latines, 1959.

Epstein, Samuel, Ralph Buchsbaum, Heinz A. Lowenstam, and Harold C. Urey. "Carbonate-Water Isotopic Temperature Scale." *Geological Society of America Bulletin* 62, no. 4 (1951): 417–26.

Epstein, Samuel, and Toshiko Mayeda. "Variation of O^{18} Content of Waters from Natural Sources." *Geochimica et Cosmochimica Acta* 4, no. 5 (1953): 213–24.

"Expéditions Polaires Françaises, Missions Paul-Emile Victor, Expéditions Arctiques: Campagne Préparatoire au Groenland, 1948 (Rapports Préliminaires 5)." Paris: Expéditions Polaires Françaises, 1954.

Farish, Matthew. "Creating Cold War Climates: The Laboratories of American Globalism." In *Environmental History and the Cold War*, edited by John R. McNeill and Corinna R. Unger, 51–84. Cambridge, UK: Cambridge University Press, 2010.

Farish, Matthew. "The Lab and the Land: Overcoming the Arctic in Cold War Alaska." *Isis* 104 (2013): 1–29.

Finsterwalder, Richard. "Expédition Glaciologique Internationale au Groënland 1957–60 (EGIG)." *Journal of Glaciology* 26, no. 3 (1959): 542–46.

Fitzhugh, William W., and Elisabeth I. Ward, eds. *Vikings: The North Atlantic Saga.* Washington, DC: Smithsonian Institution Press, 2000.

Fleming, James R. *Fixing the Sky: The Checkered History of Weather and Climate Control.* New York: Columbia University Press, 2010.

Flint, Richard F. *Glacial Geology and the Pleistocene Epoch.* New York: John Wiley, 1948.

Fogelson, Nancy. "Greenland: Strategic Base on a Northern Defense Line." *Journal of Military History* 53, no. 1 (1989): 51–63.

Fournier, Thierry. "Chapitre VI: Paul-Emile Victor et les Expéditions Américaines au Groënland, 1952–1957." Unpublished PhD thesis chapter, Ecole Nationale des Chartes, France, 2012.

Fournier, Thierry. "Chapitre VIII: Paul-Emile Victor et l'Expédition Glaciologique Internationale au Groënland, 1955–1960." Unpublished PhD thesis chapter, Ecole Nationale des Chartes, France, 2012.

Friedman, Robert Marc. "Review of Michael F. Robinson's The Coldest Crucible: Arctic Exploration and American Culture." *Isis* 99, no. 3 (2008): 641.

Frisbee, John L. "Valor: Greenland Rescue." *Air Force Magazine* 81, no. 3 (1998): 77.

Fristrup, Børge. "Recent Investigations of the Greenland Ice Cap." *Geografisk Tidsskrift* 58 (1959): 1–29.

Fristrup, Børge. "Studies of Four Glaciers in Greenland." *Geografisk Tidsskrift* 59 (1960): 89–102.

Fristrup, Børge. *The Greenland Ice Cap.* Translated by David Stoner. Copenhagen: Rhodos, 1966.
Gad, Finn. *Grønland Under Krigen.* Copenhagen: GEC Gads Forlag, 1945.
Gad, Finn. *Grønlands Historie I Indtil 1700.* Copenhagen: Nyt Nordisk Forlag Arnold Busck, 1978.
Gad, Finn. *Grønlands Historie II 1700–1782.* Copenhagen: Nyt Nordisk Forlag Arnold Busck, 1969.
Georgi, Johannes. "Greenland as a Switch for Cyclones." *Geographical Journal* 81 (1933): 344–52.
Georgi, Johannes. *Mid-Ice: The Story of the Wegener Exhibition to Greenland.* Translated by F. H. Lyon. New York: EP Dutton & Co, 1935.
Georgi, Johannes. "The First Sledge Journey Inland and the Establishment of the Eismitte Station." Translated by Winifred M. Deans. In *Greenland Journey, The Story of Wegener's German Expedition to Greenland in 1930 to 1931 as Told by Members of the Expedition and the Leader's Diary,* edited by Elsie Wegener and Fritz Loewe, 79–98. London: Blackie & Sons, 1939.
Gessain, Robert. *Un Homme Marche Devant: La Dernière Traversée du Groenland en Traineaux à Chiens.* Paris: Arthaud, 1989.
"Glaciological Conference of the American Geographical Society." *Journal of Glaciology* 1, no. 6 (1949): 336–37.
Goyette, Harold B. "DEW Line Canada-Iceland Link." *The Military Engineer* 361, Sept-Oct (1962): 325–28.
Grant, Shelagh D. *Polar Imperative: A History of Arctic Sovereignty in North America.* Vancouver: Douglas & McIntyre, 2010.
Green, Alex, and Jesse Gordon. "Slide Rules and WWII Bombing: A Personal History." *Journal of the Oughtred Society* 21, no. 2 (2012): 44–48.
Grønnow, Bjarne, and John Pind, eds. *The Paleo-Eskimo Cultures of Greenland: New Perspectives in Greenlandic Archaeology.* Copenhagen: Danish Polar Center, 1996.
Gulløv, Hans Christian, ed. *Grønlands Forhistorie.* Copenhagen: Gyldendal, 2005.
Hacquebord, Louwrens, and Dag Avango. "Settlements in an Arctic Resource Frontier Region." *Arctic Anthropology* 46 (2009): 25–39.
Haefeli, Richard. "The Development of Snow and Glacier Research in Switzerland." *Journal of Glaciology* 1, no. 4 (1948): 192–201.
Hamilton, Angus C. "Geodetic Survey of Northern Canada by SHORAN Trilateration." *Polar Record* 9, no. 61 (1959): 320–30.
Harper, Kristine C. "Climate Control: United States Weather Modification in the Cold War and Beyond." *Endeavour* 32 (2008): 20–26.
Harper, Kristine C., and Ronald E. Doel. "Environmental Diplomacy in the Cold War: Weather Control, the United States, and India, 1966–1967." In *Environmental Histories of the Cold War,* edited by John R. McNeill and Corinna R. Unger, 115–37. Cambridge: Cambridge University Press, 2010.
Heidt, Daniel. "Clenched in the JAWS of America? Canadian Sovereignty and the Joint Arctic Weather Stations, 1946–1972." *Calgary Papers in*

Military and Strategic Studies 4, Canadian Arctic Sovereignty and Security: Historical Perspectives (2011): 145–70.

Heinrich, Jens. *Eske Brun og det Moderne Grønlands Tilblivelse 1932–1964*. PhD Thesis, Ilisimatusarfik/Grønlands Universitet, 2010.

Heiskanen, Veikko A. "New Era of Geodesy." *Science* 121, no. 3133 (1955): 48–50.

Heiskanen, Veikko A. "Symposium: New Era of Geodesy." *Chronique Internationale* 3 (1954): 55–78.

Heuberger, Jean-Charles. *Expéditions Polaires Francaises, Missions Paul-Emile Victor, V: Glaciologie Groënland, Volume 1, Forages sur l'Inlandsis*. Paris: Hermann et Cie, 1954.

Heymann, Matthias. "Klimakonstruktionen: Von der Klassichen Klimatologie zur Klimaforschung." *NTM Journal of History of Sciences, Technology and Medicine* 17 (2009): 171–97.

Heymann, Matthias. "The Evolution of Climate Ideas and Knowledge." *Wiley Interdisciplinary Reviews: Climate Change* 1, no. 4 (2010): 581–97.

Heymann, Matthias, and Ronald E. Doel, eds. *Exploring Greenland: Science and Technology in Cold War Settings* (forthcoming).

Heymann, Matthias, Henrik Knudsen, Maiken L. Lolck, Henry Nielsen, Kristian Hvidtfelt Nielsen, and Christopher Jacob Ries. "Exploring Greenland: Science and Technology in Cold War Settings." *Scientia Canadensis* 33 (2010): 11–42.

Hitchcock, William I. *France Restored: Cold War Diplomacy and the Quest for Leadership in Europe, 1944–1954*. Chapel Hill, NC: University of North Carolina Press, 1998.

Hobbs, William H. "The Defense of Greenland." *Annals of the Association of American Geographers* 31, no. 2 (1941): 95–104.

Hobbs, William H. "The Greenland Glacial Anticyclone." *Journal of Meteorology* 2 (1945): 143–53.

Hobbs, William H. "Zeno and the Cartography of Greenland." *Imago Mundi* 6 (1949): 15–19.

Holbrath, Carsten. *Danish Neutrality: A Study in the Foreign Policy of a Small State*. Oxford: Clarendon, 1991.

Holtzscherer, Jean-Jacques, and Albert Bauer. "Expéditions Polaires Francaises, Missions Paul-Emile Victor, Expéditions Arctiques, Résultats Scientifiques: Contribution à la Connaissance de l'Inlandsis du Groënland, Mesures Seismiques et Synthèse Glaciologique (Communications Présentées à la Dixième Assemblée Générale de l'Union Géodesique et Géophysique Internationale Tenue à Rome en Septembre 1954)." Paris: Expéditions Polaires Francaises, 1954.

Howkins, Adrian. "A Formal End to Informal Imperialism: Environmental Nationalism, Sovereignty Disputes, and the Decline of British Interests in Argentina, 1933–1955." *British Scholar Journal* 2 (2010): 235–62.

Hummel, Laurel J. "The US Military as Geographical Agent: The Case of Cold War Alaska." *The Geographical Review* 95 (2005): 47–72.

Huntford, Roland. *Nansen: The Explorer as Hero.* Trowbridge, UK: Duckworth, 1999.
Ichac, Marcel, and Jean-Jacques Languepin. "Groënland, 20,000 Lieues sur les Glaces." Film: Shot 1949, Released 1952.
"IGY General Report no. 20, November 1963." Washington, DC: National Academies of Science.
"Images d'un Eté." Un Film des Expéditions Polaires Francaises, Missions Paul-Emile Victor. Produced by Armor-Films, 1950.
Jacobshagen, Volker, ed. *Alfred Wegener 1880–1930, Leben und Werk, Ausstellung Anlässlich der 100. Wiederkehr Seines Geburtsjahres.* Berlin: Dietrich Reimer Verlag, 1980.
Jalonen, Olli-Pekka. "The Strategic Significance of the Arctic." In *The Arctic Challenge: Nordic and Canadian Approaches to Security and Cooperation in an Emerging International Region*, edited by Kari Möttölä. Boulder, CO: Westview Press, 1988.
Jensen, Gunnar. "One Hundred Years of Crossings of Greenland's Inland Ice." *American Alpine Journal* 32 (1990): 54–65.
Jensen, Jens Arnold Diderich. *J.A.D. Jensens Indberetning om den af Ham Ledede Expedition i 1878.* Copenhagen: Meddelelser om Grønland, 1890.
Johnsen, Sigfus J. "A Fast Lightweight Core Drill." *Journal of Glaciology* 25 (1980): 169–74.
Jones, Jonathan. "Greenland's Ice Sheet Melt: A Sensational Picture of a Blunt Fact." *The Guardian*, Friday, July 27, 2012.
Jordan, Paul W. "HIRAN Instrumental Developments." *Journal of Geophysical Research* 65, no. 2 (1960): 462–66.
Joughin, Ian, Ben E. Smith, Ian M. Howat, Ted Scambos, and Twila Moon. "Greenland Flow Variability from Ice-Sheet-Wide Velocity Mapping." *Journal of Glaciology* 56, no. 197 (2010): 415–30.
"Jökull." Yearbook of the Glaciological Society of Iceland, vol. 1, 1951.
Kane, Elisha Kent. *Arctic Explorations.* Philadelphia, PA: Childs & Peterson, 1856.
Karsten, Achim, and Manfred Stober. "Deformation-Messungen auf dem Grönlandeschen Inlandeis Wahrend der Internationale Glaziologischen Grönland Expedition." *Polarforschung* 45 (1974): 45–50.
Kehrt, Christian. "Eternal Ice in the Cold War: The Polar Regions in Spatial and Environmental Perspective, 1957–1991." Paper presented at the *Cold War Science, Colonial Politics and National Identity in the Arctic Workshop*, Aarhus, December 2010.
Kehrt, Christian. "Ponies, Dogs or Propeller Sledges? Alfred Wegener and the Limits of Modern Technology in Polar Exploration." Paper presented at the SHOT Annual Meeting, Copenhagen, October 2012.
Kehrt, Christian. "The Wegener Diaries: Scientific Expeditions into the Eternal Ice." Collaborative Online Exhibition Between the Rachel Carson Center for Environment and Society and the Deutsches Museum Munich: http://www.Environmentandsociety.org/Exhibitions/Wegener-Diaries/Overview.

Kimball, Warren F. *The Juggler: Franklin Roosevelt as Wartime Statesman.* Princeton, NJ: Princeton University Press, 1991.

Kinney, D. J. "Engineering Greenland: Icecap-1 and the Militarization of Arctic Technologies." Paper presented at the SHOT Annual Meeting, Copenhagen, October 2012.

Kirwan, Laurence P. *A History of Polar Exploration.* New York: W. W. Norton & Co, 1960.

Kjems, Rud. *Horisonter af Is: Erobringen af den Grønlandske Indlandsis.* GEC Gads Forlag: Copenhagen, 1981.

Knudsen, Henrik. "Cold War, Ionospheric Research in Greenland, and the Politics of Rockets: A Study of the Ill-Fated Operation PCA 68," (forthcoming).

Knuth, Eigil. *Fire Mand og Solen: En Tur Over Grønlands Indlandsis, 1936.* Copenhagen: Gyldendal, 1937.

Koch, Johan Peter, and Alfred Wegener. *Wissenschaftliche Ergebnisse der Dänischen Expedition nach Dronning Louises-Land und Quer Über das Inlandeis von Nordgrönland 1912–1913.* Copenhagen: Meddelelser om Grønland, 1930.

Koehler, Franz A. *Special Rations for the Armed Forces, 1946–53.* Washington, DC: Office of the Quartermaster General, 1958.

Korsmo, Fae L. "The Origins and Principles of the World Data Center System." *Data Science Journal* 8 (2010): IGY55–IGY65.

Korsmo, Fae L. "Glaciology, the Arctic, and the US Military, 1945–58." In *New Spaces of Exploration: Geographies of Discovery in the 20th Century*, edited by Simon Naylor and James R. Ryan, 125–47. London: I.B. Tauris, 2010.

Korsmo, Fae L. "The Early Cold War and US Arctic Research." In *Extremes: Oceanography's Adventures at the Poles*, edited by Keith R. Benson and Helen M. Rozwadowski, 173–200. Sagamore Beach, MA: Science History Publications, 2007.

Krause, Reinhard A. "Alfred Wegener, Geowissenschaftler aus Leidenschaft: Eine Reflexion Anlässlich des 125. Geburtstages des Schöpfers der Kontinentalverschiebungstheorie." *Deutsches Schiffahrtsarkiv* 28 (2005): 299–326.

Krige, John. *American Hegemony and Postwar Reconstruction of Science in Europe.* Cambridge, MA: MIT Press, 2006.

Kuklick, Henrika, and Robert E. Kohler. "Introduction: Science in the Field." *Osiris* 11 (1996): 1–14.

Lackenbauer, P. Whitney, ed. *Canada and Arctic Sovereignty and Security: Historical Perspectives.* Calgary: University of Calgary Press—Calgary Papers in Military and Strategic Studies, 2011.

Lackenbauer, P. Whitney, and Matthew Farish. "The Cold War on Canadian Soil: Militarizing a Northern Environment." *Environmental History* 12 (2007): 920–50.

Lackenbauer, P. Whitney, Matthew Farish, and Jennifer Arnold-Lackenbauer, eds. *The Distant Early Warning (DEW) Line: A Bibliography and*

Documentary Resource List. Calgary: Arctic Institute of North America, 2005.

Lamb, Hubert Horace. "Britain's Climate in the Past (Lecture given to the British Association for the Advancement of Science at Southampton on September 2, 1964)." In *The Changing Climate: Selected Papers*, edited by Hubert Horace Lamb, 170–95. London: Methuen & Co., 1966.

Lampe, David. *Pyke: The Unknown Genius.* London: Evans Bros, 1959.

Langway, Chester C. "CRREL Report SR 31: Snow Studies and Other Observations: Operation King Dog, Sondrestrom, Greenland." US Cold Regions Research and Engineering Laboratory, 1959.

Langway, Chester C. "ERDC/CRREL Report TR-08-1: The History of Early Polar Ice Cores." US Cold Regions Research and Engineering Laboratory, 2008.

Langway, Chester C., Hans Oeschger, and Willi Dansgaard, eds. *Greenland Ice Cores: Geophysics, Geochemistry and Environment.* Washington, DC: American Geophysical Union, 1985.

Latarjet, Raymond. "Les Rations Alimentaires." *Atomes* 4 (1949): 180–82.

Launius, Roger D. "Toward the Poles: A Historiography of Scientific Exploration During the International Polar Years and the International Geophysical Year." In *Globalizing Polar Science: Reconsidering the International Polar and Geophysical Years*, edited by Roger D. Launius, James R. Fleming, and David H. DeVorkin, 47–81. New York: Palgrave Macmillan, 2010.

Launius, Roger D., James R. Fleming, and David H. DeVorkin. "Introduction: Rise of Global Scientific Inquiry in the International Polar and Geophysical Years." In *Globalizing Polar Science: Reconsidering the International Polar and Geophysical Years*, edited by Roger D. Launius, James R. Fleming, and David H. DeVorkin, 1–9. New York: Palgrave Macmillan, 2010.

Lee, Hugh J. "Peary's Transections of North Greenland, 1892–1895." *Proceedings of the American Philosophical Society* 82 (1940): 921–34.

Legget, Robert F. "Canadian Snow Conference." *Journal of Glaciology* 1, no. 3 (1948): 116–17.

LeSchack, Leonard A. "The French Polar Effort and the Expéditions Polaires Françaises." *Arctic* 17 (1964): 3–14.

Levere, Trevor H. *Science and the Canadian Arctic: A Century of Exploration, 1818–1918.* Cambridge: Cambridge University Press, 1993.

Lidegaard, Bo. *I Kongens Navn: Henrik Kauffman i Dansk Diplomati, 1919–1958.* Copenhagen: Samleren, 1996.

Livingstone, David N. *Putting Science in its Place: Geographies of Scientific Knowledge.* Chicago, IL: University of Chicago Press, 2003.

Loewe, Fritz. "The End of the Last Autumn Sledge Journey." Translated by Winifred M. Deans. In *Greenland Journey, The Story of Wegener's German Expedition to Greenland in 1930 to 1931 as Told by Members of the Expedition and the Leader's Diary*, edited by Elsie Wegener and Fritz Loewe, 170–77. London: Blackie & Sons, 1939.

Loewe, Fritz. "The Scientific Exploration of Greenland from the Norsemen to the Present, Paper presented at a Public Lecture on May 26, 1969 at Ohio State University," Columbus, Ohio: Institute of Polar Studies, Ohio State University, 1970.

Lolck, Maiken. *Klima, Kold Krig og Iskerner.* Aarhus: Aarhus Universitetforlag, 2006.

Lopez, Barry H. *Arctic Dreams: Imagination and Desire in a Northern Landscape.* New York: Charles Scribner & Sons, 1986.

Lundestad, Geir. *America, Scandinavia and the Cold War, 1945–1949.* New York: Columbia University Press, 1980.

Lyddan, Robert H. *Science in the Arctic Basin: A Report, Geodetic and Cartographic Aspects.* Washington, DC: US Committee on Polar Research, National Research Council, National Academy of Sciences, 1963.

Løkkegaard, Finn. *Det Danske Gesandtskab in Washington 1940–1942.* Copenhagen: Gyldendal, 1968.

Lüdecke, Cornelia. "Approaching the Southern Hemisphere: The German Pathway in the 19th Century." In *Globalizing Polar Science: Reconsidering the International Polar and Geophysical Years*, edited by Roger D. Launius, James R. Fleming and David H. DeVorkin, 159–75. New York: Palgrave Macmillan, 2010.

MacDonald, Robert. "Challenges and Accomplishments: A Celebration of the Arctic Institute of North America." *Arctic* 58, no. 4 (2005): 440–51.

Mahoney, E. J. "1370th Personnel Taught to Live Anywhere from Pole to Equator." *Turner Air Force Base Reflex*, 1959.

Mahoney, E. J. "1374th Fixes, Runs Electronic Equipment." *Turner Air Force Base Reflex*, 1959, 23.

Malaurie, Jean. *Hummocks 1 et 2, Collection 'Terre Humaine'.* Paris: Plom, 1999[1955].

Marret, Mario, and Fred Orani. "Nous Avons Vingt Ans." Film, Produced by La Société Nouvelle Armor-Films, 1968.

Martin, Douglas. "Willi Dansgaard Dies at 88; Read Climates in Old Ice." *The New York Times*, January 29, 2011, p. D7.

Martin, Ursula B. "Review of Alfred Wegener: The Father of Continental Drift by Martin Schwarzbach, Carla Love." *Isis* 78, no. 2 (1987): 324–25.

Martin-Nielsen, Janet. "'The Deepest and Most Rewarding Hole Ever Drilled': Ice Cores and the Cold War in Greenland." *Annals of Science* 70 (2013): 47–70.

Martin-Nielsen, Janet. "The Other Cold War: The United States and Greenland's Ice Sheet Environment, 1948–1966." *Journal of Historical Geography* 38 (2012): 69–80.

Matthes, François E. "The Glacial Anticyclone Theory Examined in the Light of Recent Meteorological Data from Greenland, Part I." *American Geophysical Union Transactions* 27 (1946): 324–41.

Maurer, John. *Local-Scale Snow Accumulation Variability on the Greenland Ice Sheet from Ground-Penetrating Radar.* MA Thesis, University of Colorado at Boulder, 2006.

McCleary, George. "Air Transportation in the Antarctic." *US Antarctic Projects Office Bulletin* 5, no. 5 (1964): 12–16.
McCook, Stuart. "'It May Be Truth, But It Is Not Evidence': Paul De Chaillu and the Legitimation of Evidence in the Field Sciences." *Osiris* 11 (1996): 177–97.
McCoy, Roger. *Ending in Ice: The Revolutionary Idea and Tragic Expedition of Alfred Wegener.* Oxford: Oxford University Press, 2006.
Miller, Edward. "Ski Hercules." *Air Classics* (January 1975): 26–29.
Mock, Steven J. "Geodetic Positions of Borehole Sites of the Greenland Ice Sheet Project." US Cold Regions Research and Engineering Laboratory, 1976.
Mock, Steven J. "Greenland Operations of the 17th Tactical Airlift Squadron and CRREL, 1973." http://www.firebirds.org/menu1/mnu1_p12.htm. Accessed October 12, 2012.
Mock, Steven J., and Donald L. Alford. "CRREL Special Report 67: Installation of Ice Movement Poles in Greenland." US Cold Regions Research and Engineering Laboratory, 1964.
Muller, Edward. "The Top of the World." *The Reader's Digest* (March 1954): 156–63.
Murphy, David Thomas. *German Exploration of the Polar World: A History, 1870–1940.* Lincoln, NE: University of Nebraska Press, 2002.
Mälzer, Hermann. *Das Nivellement Über das Grönländische Inlandeis der Internationalen Glaziologischen Grönland-Expedition 1959.* Copenhagen: Meddelelser om Grønland, 1964.
Nansen, Fridtjof. "Journey across the Inland Ice of Greenland from East to West." *Proceedings of the Royal Geographic Society and Monthly Record of Geography* 11, no. 8 (1889): 469–87.
Nansen, Fridtjof. "Journey on the Inland Ice." *Journal of the American Geographical Society of New York* 23 (1891): 171–93.
Nansen, Fridtjof. *Nord i Tåkeheimen. Utforskningen av Jordens Nordlige Strøk i Tidligere Tider.* Oslo: Kristiania, 1911.
Nansen, Fridtjof. *Paa Ski over Grønland. En Skildring af den Norske Grønlands-Ekspedition 1888–89.* Kristiania: Aschehoug, 1890.
Naylor, Simon. "The Field, the Museum and the Lecture Hall: The Spaces of Natural History in Victorian Cornwall." *Transactions of the Institute of British Geographers* 27 (2002): 494–513.
Naylor, Simon, Katrina Dean, and Martin Siegert. "The IGY and the Ice Sheet: Surveying Antarctica." *Journal of Historical Geography* 34 (2008): 574–95.
Nebeker, Frederik. *Calculating the Weather: Meteorology in the 20th Century.* San Diego, CA: Academic Press, 1995.
Needell, Allan A. *Science, Cold War and the American State: Lloyd V. Berkner and the Balance of Professional Ideals.* Amsterdam: Harwood, 2000.
Nelson, Clifford N. "Geological Survey." In *The History of Science in the United States: An Encyclopedia*, edited by Marc Rothenberg. New York and London: Garland, 2001.

Nevière, Jean. "Nivellement Géodesique de l'Indlandsis, Campagne au Groënland 1948–1949–1950." *Rapports Scientifiques des Expéditions Polaires Francaises*, 1954.

Nichols, Robert L. "Scientific Studies on the Ice Cap and in Inglefield Land with Special Reference to Military Significance, Final Report on the Scientific Program B, Operation Ice Cap 1953, Project DA 9–98–07–002." Stanford Research Institute, 1954.

Nielsen, Henry, and Henrik Knudsen. "Cold Atoms: The Hunt for Uranium in Greenland in the Late Cold War, 1970–1989." *Exploring Greenland: Science and Technology in Cold War Settings* (forthcoming).

Nielsen, Henry, and Henrik Knudsen. "Too Hot to Handle: The Controversial Hunt for Uranium in Greenland in the Early Cold War." *Centaurus* 55, no. 3 (2013): 319–43.

Nielsen, Kristian Hvidtfelt, and Christopher Jacob Ries, eds. *Scientists and Scholars in the Field: Studies in the History of Fieldwork and Expeditions*. Aarhus: Aarhus University Press, 2012.

Nijampurkar, V. N., and Henrik B. Clausen. "A Century Old Record of Lead-210 Fallout on the Greenland Ice Sheet." *Tellus* 42B (1990): 29–38.

Nuttall, Mark. "Self-Rule in Greenland: Towards the World's First Independent Inuit State?" *Indigenous Affairs* 3–4 (2008): 64–70.

Olesen, Thorsten Borring. "Tango for Thule: The Dilemmas and Limits to the 'Neither Confirm nor Deny Doctrine' in the Danish-American Relationship, 1957–1968." *Journal of Cold War Studies* 13 (2011): 116–47.

Oreskes, Naomi. *The Rejection of Continental Drift: Theory and Method in American Earth Science*. New York: Oxford University Press, 1999.

Paton, Bruce C. "Cold, Casualties, and Conquests: The Effects of Cold on Warfare." In *Medical Aspects of Harsh Environments 1*, edited by Kent B. Pandolf and Robert E. Burr. Falls Church, VI: Office of The Surgeon General, US Army, 2001.

Peary, Robert E. "Journeys in North Greenland." *The Geographical Journal* 11, no. 3 (1898): 213–39.

Peebles, Curtis. *The Corona Project: America's First Spy Satellites*. Annapolis, MD: Naval Institute Press, 1997.

Petersen, Nikolaj. "Negotiating the 1951 Greenland Defense Agreement: Theoretical and Empirical Analyses." *Scandinavian Political Studies* 21 (1998): 1–28.

Petersen, Nikolaj. "SAC at Thule: Greenland in the US Polar Strategy." *Journal of Cold War Studies* 13 (2011): 90–115.

Petersen, Nikolaj. "The Iceman That Never Came: 'Project Iceworm', the Search for a NATO Deterrent, and Denmark, 1960–62." *Scandinavian Journal of History* 33 (2008): 75–98.

Petersen, Nikolaj. "The Politics of US Military Research in Greenland in the Early Cold War." *Centaurus* 55 (2013): 294–318.

Pommier, Robert. *Au-delà de Thule: Sur la Route des Glaces*. Paris: Amiot Dumont, 1953.

Powell, Richard C. "Geographies of Science: Histories, Localities, Practices, Futures." *Progress in Human Geography* 31, no. 3 (2007): 309–29.
Powell, Richard C. "Science, Sovereignty and Nation: Canada and the Legacy of the International Geophysical Year, 1957–1958." *Journal of Historical Geography* 34 (2008): 618–638.
Prost, Antoine. *Les Origines de la Politique de la Recherche en France, 1939–1958.* Paris: Cahiers Pour l'Histoire du CNRS I, 1988.
Prost, Antoine. *Les Réformes du CNRS 1959–1960.* Paris: Cahiers Pour l'Histoire du CNRS, 1990.
Rabbitt, Mary C. "The United States Geological Survey, 1879–1989." US Geological Survey Circular 1050. US Government Printing Office, 1989.
Rack, Ursula. *Sozialhistorische Studie zur Polarforschung Anhand von Deutschen und Oesterreich-Ungarischen Polarexpeditionen Zwischen 1868–1939.* Vienna: University of Vienna, 2010.
Radok, Uwe. "The International Commission on Snow and Ice and Its Precursors, 1894–1994." *Hydrological Sciences Journal* 42, no. 2 (1997): 131–40.
Ragone, Stephen E., and C. A. Wolf. "CRREL Report No. SR 169: Analysis of the Major Cationic Constituents of the 1964 to 1969 Snow Accumulations at DYE Sites 2 and 3, Greenland." US Cold Regions Research and Engineering Laboratory, 1972.
Rasmussen, Knud. "Professor Alfred Wegener in Memoriam." *Geografisk Tidsskrift* 34, no. 2 (1931): 66–67.
Ravitts, Gail S. "Bert R.J. 'Fish' Hassell and Parker D. 'Shorty' Cramer: Pilots of a Remarkable Rockford-to-Stockholm Flight." *Aviation History*, September 2000.
Renaud, André. *Etudes Physiques et Chimiques sur la Glace de l'Indlandsis du Groënland 1959.* Copenhagen: Meddelelser om Grønland, 1969.
Renaud, André. "La Participation de la Suisse à l'Expédition Glaciologique Internationale (EGIG) de 1957 à 1961." *Revue Internationale de l'Horlogerie* (1958): U1–U4.
Reppe, Xavier. *Aurore sur l'Antarctique.* Paris: Nouvelles Editions Latines, 1957.
Reske-Nielsen, Erik, and Erik Kragh. *Atlantpagten og Danmark 1949–1972.* Copenhagen: Atlantsammenslutningen, 1972.
"Review of Mid-Ice: The Story of the Wegener Expedition to Greenland." *The Geographical Journal* 85, no. 5 (1935): 476–78.
"Review of Wissenschaftliche Ergebnisse der Deutschen Grönland-Expedition Alfred Wegener 1929 und 1930–31." *The Geographical Journal* 95, no. 5 (1940): 395–96.
Rey, Louis. *Groënland: Univers de Cristal.* Paris: Flammarion, 1974.
Ries, Christopher Jacob. "Armchairs, Dogsleds, Ships and Airplanes: Field Access, Scientific Credibility, and Geological Mapping in Northern and North-Eastern Greenland, 1900–1939." In *Scientists and Scholars in the Field: Studies in the History of Fieldwork and Expeditions*, edited by Kristian Hvidtfelt Nielsen and Christopher Jacob Ries, 329–61. Aarhus: Aarhus University Press, 2012.

Ries, Christopher Jacob. "On Frozen Ground: William E. Davies and the Military Geology of Northern Greenland, 1952–1960." *The Polar Journal* 2, no. 2 (2012): 334–57.

Riffenburgh, Beau. *The Myth of the Explorer: The Press, Sensationalism, and Geographical Discovery*. London: Belhaven, 1993.

Rink, Hinrich. "Om Isens Udbredning og Bevægelse over Nordgrønlands Fastland Samt om Isfjældenes Oprindelse." *Tidsskrift for Populære Fremstillinger af Naturvidenskaben* (1853): 103–18.

Rink, Hinrich. "The Recent Danish Explorations in Greenland and Their Significance as to Arctic Science in General." *Proceedings of the American Philosophical Society* 22, no. 120 (1885): 280–96.

Roberts, Peder. "Nordic or National? Postwar Visions of Polar Conflict and Cooperation." In *Science, Geopolitics and Culture in the Polar Regions: Norden Beyond Borders*, edited by Sverker Sörlin, 55–78. Surrey, England: Ashgate, 2013.

Roberts, Peder. "Science and Political Authority in the Polar Regions in the Cold War – and Beyond." Talk given at Aarhus University, February 25, 2013.

Roberts, Peder. *The European Antarctic: Science and Strategy in Scandinavia and the British Empire*. New York: Palgrave Macmillan, 2011.

Robinson, Michael F. *The Coldest Crucible: Arctic Exploration and American Culture*. Chicago, IL: University of Chicago Press, 2006.

Ross, Frederic S., and Paul E. Ancker. "Thule Air Base." *Tidsskriftet Grønland* 9–10 (1977): 268–78.

Ross, John E. R. "Geodetic Observations in the Canadian Arctic." *Arctic* 7 (1954): 191–94.

Russell, Richard J. "Instability of Sea Ice." *American Scientist* 45, no. 5 (December, 1957): 414–30.

Ryan, James R., and Simon Naylor. "Exploration and the 20th Century." In *New Spaces of Exploration: Geographies of Discovery in the 20th Century*, edited by Simon Naylor and James R. Ryan, 1–22. London: I.B. Tauris & Co., 2010.

Saladin d'Anglure, Bernard. "Mauss et l'Anthropologie des Inuit." *Sociologie et Sociétés* 36, no. 2 (2004): 91–130.

Sapp, Bill, and Gordon Barnes. "1374th Mapping and Charting Squadron: A Tribute to Those Ground Station Guys." http://www.1370th.org/1370pmw/1374/Tribute.htm.

Scavenius, Erik. *Forhandlingspolitiken Under Besættelsen*. Copenhagen: Steen Hasselbalchs Forlag, 1947.

Schwarzbach, Martin. *Alfred Wegener und die Drift der Kontinente*. Stuttgart: Wissenschaftliche Verlagsgesellschaft, 1986.

Seaver, Kirsten A. "Review of Narrating the Arctic: A Cultural History of Nordic Scientific Practices." *Arctic* 56 (2003): 305–06.

Seaver, Kirsten A. *The Frozen Echo: Greenland and the Exploration of North America, ca. A.D. 1000–1500*. Stanford, CA: Stanford University Press, 1996.

Seeley, Stuart William. "SHORAN: A Precision Five Hundred Meter Yardstick." *Proceedings of the American Philosophical Society* 105, no. 4 (1961): 447–51.
Seligman, G. "Meeting of the International Commission on Snow and Glaciers, Oslo." *Journal of Glaciology* 1, no. 5 (1948): 289–92.
Sennert, H.G. "The History of SHORAN." *Photogrammetric Engineering* 12, no. 4 (1946): 375–77.
Simpson, C. James W. "The British North Greenland Expedition." *The Geographical Journal* 121, no. 3 (1955): 274–89.
Skrotzky, Nicolas. *Terres Extrêmes: La Grande Aventure des Pôles.* Paris: Editions Denoël, 1986.
Smith, James Raymond. *Introduction to Geodesy: The History and Concepts of Modern Geodesy.* New York: John Wiley & Sons, 1997.
Solovey, Mark. *Shaky Foundations: The Politics-Patronage-Social Science Nexus in Cold War America.* New Brunswick, NJ: Rutgers University Press, 2013.
Sorge, Ernst. "Glaziologische Untersuchungen in Eismitte." In *Wissenschaftliche Ergebnisse der Deutschen Groenland Expedition Alfred Wegener 1929 und 1930–31.* Leipzig: F.A. Brokaus, 1935.
Sorge, Ernst. "Scientific Results of the Wegener Expedition to Greenland." *Geographical Journal* 81 (1933): 333–34.
Sorge, Ernst. "Winter at Eismitte." Translated by Winifred M. Deans. In *Greenland Journey, The Story of Wegener's German Expedition to Greenland in 1930 to 1931 as Told by Members of the Expedition and the Leader's Diary*, edited by Elsie Wegener and Fritz Loewe, 179–97. London: Blackie & Sons, 1939.
Spender, Michael, and Therkel Mathiassen. "Alfred Wegener's Greenland Expeditions 1929 and 1930–31: Review." *The Geographical Journal* 84 (1934): 515–22.
Spufford, Francis. *I May Be Some Time: Ice and the English Imagination.* New York: St. Martin's Press, 1997.
Stefansson, Vilhjalmur. *The Friendly Arctic: The Story of Five Years in Polar Regions.* New York: Macmillan Co., 1921.
Storgaard, Einar. "Alfred Wegeners Grønlandsekspedition 1929–1931." *Geografisk Tidsskrift* 35, no. 4 (1932): 198–214.
Stäblein, G. "Alfred Wegener, From Research in Greenland to Plate Tectonics." *GeoJournal* 7, no. 4 (1983): 361–68.
Suid, Lawrence H. *The Army's Nuclear Power Program: The Evolution of a Support Agency.* Westport, CT: Greenwood Press, 1990.
Suzuki, Toshitaka, Akira Maki, Kazutake Ohta, Kokichi Kamiyama, Yoshiyuki Fujii, and Okitsugu Watanabe. "Determination of Lead-210 in an Ice Core from an Arctic Glacier by an Alpha-Ray Measurement." *Analytical Sciences* 12 (1996): 923–26.
Sörlin, Sverker. "Narratives and Counter Narratives of Climate Change: North Atlantic Glaciology and Meteorology, c. 1930–1955." *Journal of Historical Geography* 35 (2009): 237–55.

Sörlin, Sverker. "The Global Warming That Did Not Happen: Historicizing Glaciology and Climate Change." In *Nature's End: Environment and History*, edited by Sverker Sörlin and Paul Warde, 93–114. London: Palgrave Macmillan, 2009.

Taagholt, Jørgen, and Jens Claus Hansen. *Den Nye Sikkerhed: Grønland i et Sikkerhedspolitisk Perspektiv.* Denmark: Atlantsammenslutningen, 1999.

Teal, John J. "Greenland and the World Around." *Foreign Affairs* (October, 1952): 128–41.

Teissier, Pierre. "Solid-State Chemistry in France: Structures and Dynamics of a Scientific Community since World War II." *Historical Studies in the Natural Sciences* 40 (2010): 225–58.

"The Greenland Ice Plateau." *Air University Quarterly Review*, 1955.

"The Non-Existence of Peary Channel." *Geographical Review* 1, no. 6 (1916): 448–52.

"The Saga of Erik the Red," English translation of the original 'Eiríks Saga Rauða' by John Sephton, 1880.

"The Soviet High Latitude Arctic Air Expedition and Drifting Stations in the Arctic Ocean." *Polar Record* 7 (1955): 396–99.

Theberge, Albert E. "The Making of an Earth Measurer." *Hydro International* 13, no. 7 (2009): 3.

Tschaen, Louis. "Groënland 1948–1949–1950: Astronomie-Nivellement Géodesique sur l'Indlandsis, Nouveau Calcul." Paris: Expéditions Polaires Francaises, 1959.

Tschaen, Louis, and Albert Bauer. "Le Mouvement de la Partie Centrale de l'Inlandsis du Groënland." Association Internationale des Sciences Hydrologiques, 1958.

Turchetti, Simone, Simon K. Naylor, Katrina Dean, and Martin Siegert. "On Thick Ice: Scientific Internationalism and Antarctic Affairs, 1957–1980." *History and Technology* 24 (2008): 351–76.

"Twentieth Anniversary of Expéditions Polaires Francaises (Missions Paul-Emile Victor)." *Arctic* 21 (1968): 59–66.

Urey, Harold C., Heinz A. Lowenstam, Samuel Epstein, and C. R. McKinney. "Measurement of Paleotemperatures and Temperatures of the Upper Cretaceous of England, Denmark and the Southeastern United States." *Geological Society of America Bulletin* 62, no. 4 (1951): 399–416.

van der Veen, Cornelis J., Kenneth C. Jezek, Ellen Mosley-Thompson, Ian M. Whillans, and John F. Bolzan. "Two Decades of Glaciological Investigations in South and Central Greenland." *Polar Geography* 24 (2000): 259–349.

Vetter, Jeremy, ed. *Knowing Global Environments: New Historical Perspectives on the Field Sciences.* New Brunswick. NJ: Rutgers University Press, 2010.

Victor, Paul-Emile. *Apoutsiak, Le Petit Flocon de Neige.* Paris: Flammarion, 1948.

Victor, Paul-Emile. *Banquise: Le Jours Sans Ombre.* Paris: Grasset, 1939.

Victor, Paul-Emile. *Boréal: La Joie dans la Nuit.* Paris: Grasset, 1938.

Victor, Paul-Emile. "Expéditions Polaires Francaises." *Atomes* 4 (1949): 121–32.

Victor, Paul-Emile. "Impressions du Bout du Monde." *Marianne*, February 2, 1938.
Victor, Paul-Emile. "Je suis un Esquimau." *Paris Soir*, December 29, 1937.
Victor, Paul-Emile. *L'Iglou*. Paris: Stock, 1987.
Victor, Paul-Emile. "Les Jeux de Ficelle Chez les Eskimos d'Angmagssalik." *Journal de la Société des Américanistes* 29 (1937): 387–95.
Victor, Paul-Emile. *Mes Aventures Polaires*. Paris: Editions GP, 1975.
Victor, Paul-Emile. *My Eskimo Life*. London: Hamish Hamilton, 1939.
Victor, Paul-Emile. "The French Expedition to Greenland, 1948." *Arctic* 2 (1949): 135–48.
Victor, Paul-Emile, Jean-Jacques Languepin, Marcel Ichac, and Jacques Masson. *Groënland: 1948–1949*. Paris: Arthaud, 1951.
Vidal, Joseph. *Mesures Effectuées par le Groupe de Géodesie B le Long du Profil Transversal Est-Ouest de l'Indlandsis du Groënland, EGIG, Eté 1967*. Copenhagen: Meddelelser om Grønland, 1983.
Villaume, Poul. *Allieret Med Forbehold: Danmark, NATO og den Kolde Krig, Et Studie i Dansk Sikkerhedspolitik, 1949–1961*. Copenhagen: Eirene, 1995.
Villaume, Poul, and Thorsten Borring Olesen. *I Blokopdelingens Tegn, 1945–1972*. Copenhagen: Gyldendaal, 2005.
Voss, Jutta. "Alfred Wegeners Weg als Polarforscher." In *125 Jahre Deutsche Polarforschung*, edited by Alfred Wegener Institut, 81–94. Bremerhaven: Alfred-Wegener-Institut für Polar- und Meeresforschung, 1994.
Wager, Walter. *Camp Century: City Under the Ice*. Philadelphia, PA: Chilton Books, 1962.
Wallerstein, George. "Movement Observations on the Greenland Ice Cap." US Snow Ice and Permafrost Research Establishment, 1958.
Wallerstein, George. "Refraction Observations on the Greenland Icecap." *Navigation* 5, no. 3 (1956): 138–40.
Walls, J. Kermit. "The RC-130A Aircraft: A New World Mapping System." *Photogrammetric Engineering* (June 1960): 395–401.
Warner, Deborah J. "From Tallahassee to Timbuktu: Cold War Efforts to Measure Intercontinental Distances." *Historical Studies in the Physical and Biological Sciences* 30 (2000): 393–415.
Warner, Deborah J. "Political Geodesy: The Army, the Air Force and the World Geodetic System of 1960." *Annals of Science* 59 (2002): 363–89.
Washburne, Norman F. "A Survey of Human Factors in Military Performance in Extreme Cold Weather, Report AD 477889." The George Washington University Human Resources Research Office/Department of the Army, 1960.
Weart, Spencer. *The Discovery of Global Warming*. Cambridge, MA: Harvard University Press, 2008.
Wegener, Alfred. *Die Entstehung der Kontinente und Ozeane*. Braunschweig: F. Vieweg, 1915.
Wegener, Alfred. *Mit Motorboot und Schlitten in Grönland*. Bielefeld and Leipzig: Verlag von Velhagen Klasing, 1930.

Wegener, Alfred. "Tagebücher, April 1930-September 1930." Deutsches Museum Archive Nl001/014, available online at http://www.Environ mentandsociety.org/Exhibitions/Wegener-Diaries/Overview.
Wegener, Alfred. "Tagebücher, June 1906-August 1908." Deutsches Museum Archive Nl001/005, available online at http://www.Environmentandsociety.org/Exhibitions/Wegener-Diaries/Overview.
Wegener, Alfred, and Johan P. Koch. *Durch die Weiße Wüste; Die Dänische Forschungsreise Quer Durch Nordgrönland 1912–1913*. Berlin: Julius Springer, 1919.
Wegener, Elsie, and Fritz Loewe, eds. *Greenland Journey, the Story of Wegener's German Expedition to Greenland in 1930 to 1931 as Told by Members of the Expedition and the Leader's Diary*. London: Blackie & Sons, 1939.
Wegener, Kurt. *Wissenschaftliche Ergebnisse der Deutschen Grönland-Expedition Alfred Wegener 1929 und 1930/1931, Band I: Geschichte der Expedition*. Leipzig: F.A. Brockhaus, 1933.
"Wegeners Grønlands Expedition." *Geografisk Tidsskrift* 32, no. 1 (1929): 240.
Weigert, Hans W. "Iceland, Greenland and the United States." *Foreign Affairs* 23, no. 1 (1944): 112–22.
Weiss, Erik D. "Cold War Under the Ice: The Army's Bid for a Long-Range Nuclear Role, 1959–1963." *Journal of Cold War Studies* 3 (2001): 31–58.
Whayne, Tom. *Cold Injury in World War II: A Study in the Epidemiology of Trauma*. PhD Thesis, Harvard School of Public Health, 1950.
Wilkes, Owen, and Jan Øberg. *Military Research and Development in Denmark and Greenland*. Lund, Sweden: Lund University, 1982.
Williams, Lee R. "Final Report, Project 53-ASF-1, Phase 1." Headquarters, 55th Strategic Reconnaissance Wing Detachment #2, US Air Force, 1953.
Wilson, Eric G. *The Spiritual History of Ice: Romanticism, Science, and the Imagination*. New York: Palgrave Macmillan, 2003.
Wråkberg, Urban. "The Politics of Naming: Contested Observations and the Shaping of Geographical Knowledge." In *Narrating the Arctic: A Cultural History of Nordic Scientific Practices*, edited by Michael Bravo and Sverker Sörlin, 155–98. Canton, MA: Science History Publications, 2002.
Wutzke, Ulrich. *Durch die Weisse Wüste: Leben und Leistungen des Grönlandforschers und Entdeckers der Kontinentaldrift Alfred Wegener*. Gotha: Perthes, 1997.
Zallen, Doris T. "Louis Rapkine and the Restoration of French Science after the Second World War." *French Historical Studies* 17 (1991): 6–37.

Index

Aero Service Corporation, 172n115
aerology, 23
Agreement Relating to the Defense
    of Greenland (1941), 19
Ahlmann, Hans W., 22, 26, 34,
    52–3, 54, 86, 94, 117–18
air drops, 48–9, 51, 70–5, 90, 91,
    103, 108, 109, 152n56, 154n74
aircraft. *See* aviation
Alaska, 40, 41, 47, 52, 63, 64, 68,
    77, 85, 101, 112
Alford, Donald L., 106–7, 109, 110
Andersen, Einar, 97
Angmagssalik (now Ammassalik), 40
Antarctica, 2, 5, 7, 21, 39, 42, 43,
    44–5, 47, 52, 59, 109, 110–11,
    115, 120, 146n93, 175n145,
    175n146
anthropology, 40–1
anticyclones, 135n106
Arctic, 9, 12, 14, 18, 20, 39, 42–3,
    44, 47, 48, 54, 57, 61–3, 85–6,
    94, 102, 105
    exploration, 1–3, 5, 6–8, 13, 22,
        24–5, 30, 40, 42, 57–9, 82,
        108–9, 111–12, 114, 116
    See also *individual countries and
        regions*
Arctic, Desert and Tropic
    Information Center (ADTIC),
    41, 73
Arctic Institute of North America
    (AINA), 62, 63, 85–6

Aslakson, Carl I., 101–2
atmospheric contamination, 2, 9,
    92, 95, 120, 122
    Russian H-bomb tests, 9, 95
Auriol, Vincent, 43
Austria, 36, 65, 86, 87, 89
aviation, 42–4
    in Antarctica, 109, 110–11
    C-47, 62, 109–10
    C-130 Hercules, 110–11, 113,
        121, 174n132, 174n135
    early polar aviation, 3, 7, 29, 57,
        109, 140n42
    in Greenland (*including* opening
        of), 19, 29, 34–5, 65, 66, 79,
        106–7, 109–14
    ice sheet challenges/dangers, 29,
        73, 74, 109–11, 173n121
    jet assistance, 110, 111
    Lockheed, 82, 110
    at/near the North Pole, 29, 43,
        104, 110
    Project Snowman, 66, 109–10
    and the United States Air Force,
        109–10
    and the United States Navy,
        109–10
    unprepared snow landings,
        109–11, 173n122
    See also air drops; *individual
        expeditions – technological
        vision; individual expeditions –
        transportation technologies*

INDEX

Bader, Henri, 35–6, 65, 68–9, 70, 75, 76, 86, 89–90, 91, 105, 136n113, 178n26
Balchen, Bernt, 19
Base Dumont d'Urville. *See* Terre Adélie
Bauer, Albert, 49, 53, 76, 88, 89, 90, 91, 97, 108, 158–9n7
Beaudry, Emil, 61–2, 109–10, 148n8, 174n128
Benson, Carl S., 9, 64, 66–77, 78, 81–3, 91, 106
Bornholm, 125n18
Bouché, Michel, 45, 51
Bounds, R.G., 105
British Arctic Air Route Expedition (1930–1931), 133n69
British-Norwegian-Swedish Antarctic Expedition, 52
Brockamp, Bernhard, 36, 158–9n7
Brun, Eske, 55–6, 80, 92, 97–100, 117, 119–22
Bulganin, Marshall N.A., 130–1n47
Bush, Vannevar, 65
Byrd, Richard, 29, 150n22

Camp Century, 66, 82, 96, 121, 150n24, 157n103, 174n132, 179–80n36
Canada, 7, 17, 29, 40, 43, 57, 62, 69, 70, 77, 86, 101–2, 104, 112, 121, 149n14, 177n12
cartography, 68, 97, 104, 116, 117
Charcot, Jean, 45
Christie, Robert W., 68, 73, 154n74
climate change. *See* climatology/ climate
climatology/climate, 2, 3, 4, 9, 34, 36, 52, 54, 59, 65, 77–8, 87, 92–6, 106, 119, 120–22
Coe, Robert, 81
Cold Regions Research and Engineering Laboratory (CRREL), 63, 82, 106
Cold War, 6, 90, 115
and geodesy, 100–2

and Greenland, strategic significance of, 2, 9, 19–20, 61–3, 79, 83, 112
and Greenland, United States presence in, 2, 5, 9, 19–20, 61–7, 70–83
superpower conflict, 7, 20, 61–2, 101, 112
United States polar strategy (*including* science-based), 2, 5, 61–6, 82–3
See also Denmark: and the Cold War; Distant Early Warning (DEW) Line; Greenland, history of: and the Cold War; missiles, long range; North American continental security/defense
Commission for Scientific Research in Greenland (*Kommissionen for Videnskabelige Undersøgelser i Grønland*), 55, 143n72
Courtauld, Augustine, 133n69
Cramer, Parker, 109
crevasses, 8, 11, 13–14, 16, 27, 46, 56–7, 66, 67, 73

Dalager, Lars, 8, 14, 18
Danish West Indies, 19, 129n40
Danmark expedition (1906–1908), 23, 118, 132n61
Dansgaard, Willi, 78, 89, 95–6, 113–14, 120–1
de Caunes, Georges, 59
de Gaulle, Charles, 88, 138n23
de Quervain, Marcel, 93
de Veau Blosseville, Lieutenant, 45
Defense of Greenland Agreement (1951), 20, 67, 79, 82, 117, 130–1n47, 156n87
Denmark, 37, 43
Army Geographical Service, 98
and the Cold War, 5, 19–20, 78, 130–1n47
control of scientific work in Greenland, 5, 9, 19, 25, 54–6, 80–2, 89, 96–100, 104,

115–22, 143n72, 172n118, 178n26
  and Eismitte, 5, 115, 118, 120
  and the Federal Republic of Germany, 158–9n7
  Foreign Ministry, 8, 55, 80, 97–8, 104, 115
Frederik IV, 13, 127n15
Frederik VI, 115
Geodetic Institute (*Geodætisk Institut*), 98, 117
German occupation of, 5, 19, 78, 97, 98, 155n79
and glaciology (*including* ice cores), 9, 65, 78, 82, 88, 89, 95–6, 97, 98, 100, 113–14, 118–22, 178n21, 179–80n36
and the International Geophysical Year (IGY), 118–19
Meteorological Institute (*Meteorologiske Institut*), 117
metropolitan Denmark. *See* metropolitan Denmark
Ministry of Defense, 80
NATO membership, 20, 79, 98, 130–1n47
  and Norway, 5, 18, 115–16, 118, 125n17, 127n15, 176n5
  Schleswig-Holstein, 158–9n7
  and scientific nationalism, 54–5, 97–8, 115–18, 122
  sovereignty over Greenland, 5, 18, 19, 23, 54–6, 78–82, 97, 115–17, 119–20, 130–1n47
  and the Soviet Union, 130–1n47
  and the United States, 19–20, 78–82, 97, 104, 117
  and World War II, 19, 78. *See also* Denmark: German occupation of; Greenland; Greenland, history of; *individual expeditions*
Devers, Jacob L., 63
Devold, Hallvard, 116
Disko Bay, 46, 92
Distant Early Warning (DEW) Line, 112–14, 121, 174n138

d'Orléans, Duc, 45
Dubois, Georges, 53
Dumont d'Urville, Jules Sébastian César, 43, 45
DYE Stations. *See* Distant Early Warning (DEW) Line

Egede, Hans, 127n14
Eismitte
  appeal in the scientific imagination, 1–2, 3, 5–6, 8, 13, 25, 37, 39, 57, 92, 108, 114
  Danish absence, 5, 115, 118, 120
  decline in the scientific imagination, 7, 9, 108, 111–14
  expeditions. *See individual expeditions*
  location, xvii–xxi, 1, 25
  movement over time, 106–7
  scientific activity. *See individual expeditions*
  scientific record, 3, 4, 53, 70, 92, 122
  scientific stations, 4. *See also individual expeditions*
Epstein, Samuel, 77–8, 95
Erik the Red, 17
Expédition Glaciologique Internationale au Groënland (EGIG),
  and Austria, 86, 87, 89
  climate/environmental work, 2, 9, 87, 92–6
  conception and planning, 86–92
  and Denmark, 5, 9, 86, 88, 89, 95–100, 115, 118, 119–20, 158–9n7
  division of labor, 87–9, 91
EGIG II, 172n118
  and Eismitte, 1, 2, 87, 92–5, 112
  and Expéditions Polaires Françaises (EPF), 87, 90–1, 97
  and the Federal Republic of Germany, 86, 87, 89, 92, 158–9n7

Expédition Glaciologique—*Continued*
and France, 86, 87–90, 92, 99, 100, 160n17
funding, 88–9
ice core research, 9, 87, 94–6, 121
inter-country relations, 87–9, 96–100, 115, 119
and the International Geophysical Year (IGY), 87, 89–90
Jarl-Joset Station, 87, 90
permission to work in Greenland, 88, 96–9, 118
personnel. *See individual names*
publicity, 100
route, xx, 87, 92
scientific work, 87–9, 91–6, 98–9, 106, 107
and Switzerland, 86, 87, 88, 89, 93, 94–5
transportation technologies, 89, 90–1
and the United States, 88–91, 97–8, 160n17
Expéditions Polaires Françaises (EPF), 69
and Antarctica, 42–3, 44–5, 59, 90, 146n95
and Denmark, 96, 100, 120
and Expédition Glaciologique Internationale au Groënland (EGIG), 87, 90–1, 97
founding, 42, 43–4
funding, 6, 42–5
growth, 9, 59
headquarters, 51
and the International Geophysical Year (IGY), 90
mandate, 42, 44
and the Soviet Union, 146n93
and the United States, 44, 73, 75, 146n93
See also Expéditions Polaires Françaises (EPF) expedition to Greenland, 1948–1953; France; Victor, Paul-Emile

Expéditions Polaires Françaises (EPF) expedition to Greenland, 1948–1953
conception and planning, 39, 46–7
deaths, 56–7
and Denmark, 54–7, 143–4n73
division of labor, 57, 59
and Eismitte, xviii, 1, 2, 8, 45, 46–8, 53, 57, 111–12
food, 44, 50–1
funding, 6, 42–5
ice core research, 52
narratives, 8–9, 42, 57–9
overwinter living conditions, 48, 49–51
permission to work in Greenland, 54–6
personnel. *See individual names*
promotion/publicity, 8, 44, 48, 57–9
reception/legacy, 53–4, 58–9, 70, 87
scientific work, 2, 45–6, 48, 51–4, 57, 92, 106, 107
Station Centrale, 2, 45–6, 48–52, 56, 70, 75–6, 92, 102, 118
technological vision, 42, 48, 57, 141n52
transportation technologies, 8, 46–9, 57, 73
See also Expéditions Polaires Françaises (EPF); France
Exploration. *See* Arctic: exploration

Faroe Islands, 115, 125n17, 125n18
Feaver, Herbert F., 121
Finland, 41
Finsterwalder, Richard, 91
Flint, Richard F., 53, 64
Foulkes, Charles, 62
France, 5, 41, 57–8, 65, 99–100
Antarctic claim, 5, 7, 39, 44, 59, 115. *See also* France: Terre Adélie
Armée de l'Air, 44, 89, 98, 160n17
early polar exploration, 39, 40–1, 43, 45

INDEX

Fourth Republic, 6
French Southern and Antarctic Lands (*Terres Australes et Antarctiques Françaises*), 146n95
Geodetic Institute, 99
Hydrotechnical Society, 85
Institute of Food Hygiene and Stewardship, 50
National Center for Scientific Research (*Centre National de la Recherche Scientifique*), 44, 88–9, 146n95
polar expeditions, government/diplomatic support for, 6, 42–5, 55, 58, 88–9
science post-World War II, 44–5, 88–9
Terre Adélie, 42–3, 45, 59, 90, 137–8n16
See also individual expeditions; individual names
Frederikshaab (now Paamiut), 14
Frederikshaab Glacier, 11, 14
French Polar Expeditions. *See* Expéditions Polaires Françaises
Freuchen, Peter, 25
Fristrup, Børge, 58–9, 64, 82, 91, 95, 98–100, 105, 107, 113, 118–19, 121, 158–9n7

Gates, Robert W., 104
geodesy, 2, 3, 9, 88, 89, 91, 92, 98–9, 100–8, 112, 113, 172n118. *See also* HIRAN; SHORAN; World Geodetic System 1960 (WGS 60)
Geodetic Institute, Denmark (*Geodætisk Institut*), 98, 117
geodetic systems. *See* geodesy
Georgi, Johannes, 26–35, 37, 45, 47, 51, 92
Germany, 1, 8, 36, 179–80n36
early polar explorations, 20–1
Emergency Association of Germany Science (*Notgemeinschaft der Deutschen Wissenschaft*), 21, 22, 132n57
return to the polar world post-World War I, 7, 21–2, 24–25
return to the polar world post-World War II, 87, 158–9n7
and World War II, 5, 19, 44, 78, 98, 100, 155n79
See also individual expeditions
Gessain, Robert, 40
Gillis, James E., 105
glacial anticyclones. *See* anticyclones
glaciology, 3, 9, 16, 25, 33, 35–6, 37, 48, 49, 51–3, 63, 64–7, 69, 70, 76–8, 85–100, 103, 106, 113–14, 118–22, 146n93, 179–80n36
See also ice cores
Godthaab (now Nuuk), 13–14, 15, 16, 127n14
Goldberg, Edward D., 95
Goldthwait, Richard P., 65, 150n22
Greenland
and aviation. (*see* aviation: in Greenland (*including* opening of))
Commission for Scientific Research in Greenland (*Kommissionen for Videnskabelige Undersøgelser i Grønland*), 55, 143n72
Danish infrastructure/logistics, 120–1
Danish scientific activity, 5, 19, 45, 54–5, 78, 82, 88, 89, 95–6, 97–8, 99–100, 113–14, 115–22, 179–80n36
early exploration and settlement, xvii, 4–5, 8, 11–18, 23–5, 29, 40–1, 45, 116, 118, 128n23, 133n69
Geological Survey (*Grønlands Geologiske Undersøgelser*), 117
mapping, 14, 18, 23, 25, 37, 98–9, 104, 117, 132n61
natural resources, 17–18, 19, 79, 96, 155n80

Greenland—*Continued*
Office/Ministry (*Grønlands Styrelse/Ministeriet*), 25, 55, 80, 97, 99, 115, 117, 119, 122, 133n72, 143–4n73, 156n89
permission for foreign scientific work, 25, 54–6, 80–2, 96–100, 104, 118, 119–20
as a strategic space, 2, 5, 9, 17–20, 34–5, 61–2, 63, 78–9, 83, 102
Three Year Expedition to East Greenland (1931–1934), 116
United States defense areas, 20, 80–1, 97–8, 119
United States military facilities, xix, 5, 9, 19–20, 54, 62, 66, 70, 80, 82, 97–8, 102, 112–14, 121, 150n24. (*see also individual air bases and military facilities*)
United States weather stations, 19, 35, 61, 117
See also Denmark; Greenland, geography of; Greenland, history of; *under individual names*; *under individual place names*
Greenland, geography of, 4, 11–13, 18, 62–3
ice coverage, extent of, 8, 11–12, 15–16, 17, 25, 37
ice melt, 2, 37, 53, 64, 106, 120, 122
ice movement/flow, 52–3, 64, 66, 92, 93, 103, 106–8. (*see also* snow markers)
ice sheet, 1, 2–3, 8, 11–13, 14–15, 23, 24–5, 36–7, 46, 49, 53, 73–4, 105, 106–8, 111–12, 114, 120
ice thickness, 31, 34, 36–7, 52–4, 75, 120
shape, 3, 16, 17–18, 36–7, 53–4, 122, 132n61
size, 4, 13
snowfall/accumulation, 2, 32, 36, 37, 76, 77, 106, 113, 123n3, 171n112
See also crevasses

Greenland, history of
Agreement Relating to the Defense of Greenland (1941), 19
and the Cold War, 2, 5, 9, 19–20, 61–3, 66, 78–9, 82–3, 112, 130–1n47
Danish colonization, 12, 13–14, 18, 98–9, 116, 127n14, 175n1
Danish-Norwegian disputes, 5, 18, 115–16, 125n17, 176n5
Danization, 175n1
Defense of Greenland Agreement (1951), 20, 67, 79, 82, 117, 130–1n47, 156n87
Denmark-Norway, 18, 115, 125n17, 127n15
Distant Early Warning (DEW) Line, 112–14, 121, 174n138
early exploration and settlement, xvii, 4–5, 8, 11–18, 23–5, 29, 40–1, 45, 116, 118, 128n23, 133n69
East Greenland, 115–16, 118, 125n17
Erik the Red's Land (*Eirik Raudes Land*), 116
first crossing (1888), xvii, 4, 15–16
Home Rule, 175n1
Hope Colony (*Haabets Koloni*), 127n14
International Court of Justice ruling, 5, 18, 116, 125n17
migration, 17
Napoleonic Wars, 115
and NATO, 5, 20, 62, 63, 79, 117
Norse settlements, 13–14, 17–18, 127n18
Norwegian dependency, 18
Norwegian occupation (1931), 116, 125n17
Østerbygd (Eastern Settlement), 14, 127n18
Paleo-Eskimo cultures, 17, 128n33
prehistory, 17
self-determination, 175n1

INDEX

Thule Air Base, xix, 20, 66, 67, 68–9, 70–1, 74, 75, 76, 80, 81, 82, 99, 105, 110, 121, 175–6n2
Treaty of Kiel, 125n17
United States interest and activity, 2, 5, 9, 18–20, 35, 54, 61–70, 78–83, 90, 97, 103, 105–7, 109–10, 119, 171n109
Uummannaq, 99
during World War II, 5, 19, 34–5, 54, 78, 98–9
See also Denmark; Greenland; individual names; individual place names
Greenland expedition of 1912–1913 (Koch-Wegener), xvii, 16–17
Greenland Expeditions of the University of Michigan (1926–1931), 109, 133n69
Greenland Ice Sheet Project (GISP), 113–14, 174n132, 179–80n36
Greenlanders. See Kalaallit
Guillard, Robert, 49

Haefeli, Richard, 88, 91, 97, 106
Hansen, Hans C., 130–1n47
Hassell, Bert, 109, 173n123
Helk, Jorgen, 97–8
HIRAN
 accuracy, 102
 and Air Photographic and Charting Service, 103, 105
 and Denmark, 104
 and Eismitte (Station 31), xxi, 102–7, 109, 112, 171n106
 and Greenland, xxi, 102–7, 110, 169n93
 North Atlantic tie, 102, 104
 origins, 101–2
 permission to work in Greenland, 104
 secrecy, 107, 172n115
 and SIPRE, 103–6
 technique, 101–3
 transportation technologies, 110

See also Aslakson, Carl I.; geodesy; SHORAN
Hobbs, William H., 133n69
Holston, James B., 68
Holtzscherer, Jean-Jacques, 49, 53
Hough, Floyd W., 100

Ice Age, 15, 54, 96, 121
ice cores, 2, 3, 9, 36, 51–2, 66, 67, 76–8, 85, 87, 89, 90, 93–6, 113–14, 120–2, 174n132, 175n145, 179–80n36
ice melt, 2, 37, 53, 64, 106, 120, 122
ice sheet. See Greenland, geography of: ice sheet
Iceland, 17, 26, 49, 56, 59, 85, 95, 102, 104, 112, 115, 125n17, 179–80n36
Icelandic Glaciological Society, 85
Indlandsisen. See Greenland, geography of: ice sheet
International Commission of Snow and Ice, 86, 88
International Geophysical Year (IGY), 77, 87, 88–90, 110, 118–19
International Glaciological Expedition to Greenland. See Expédition Glaciologique Internationale au Groënland
International Glaciological Society, 86
International Polar Years, 45, 90

Jarl, Jens, 55–7, 87
Jensen, Jens Arnold Diderich, xvii, 11, 14, 126n1
jet streams, 26, 133n75
Joset, Alain, 56–7, 87
Juneau Ice Field Research Project (US), 52

Kalaallit, 4, 5, 12, 14, 17, 27–8, 55, 80, 119, 122, 156n89, 175–6n2
Kane, Elisha Kent, 12
Kauffman, Henrik, 19, 98–9
Kirwan, Laurence P., 11, 24
Knuth, Eigil, 40, 55, 117, 118

Koch, Johan Peter, 24, 25
  Danmark expedition
    (1906–1908), 23
  Greenland Expedition
    (1912–1913), xvii, 16–17
Koch, Lauge, 29, 45, 54, 116
Köppen, Wladimir, 24

Lamb, Hubert Horace, 93–4
Landsberg, Helmut E., 94
Langway, Chester C., 77, 113–14, 179n30
Larsen, Helge, 25, 89, 91, 97–8
Lindbergh, Charles, 29
Loewe, Fritz, 26–30, 34, 37, 47, 51, 111, 114
Loubry, Roger, 49

mapping, 1, 2, 29, 68, 124n8.
  See also cartography; geodesy;
  Greenland: mapping
Marret, Mario, 50, 57–8, 141n52
Matter-Steveniers, Fred, 40
Matthes, François E., 34–5
Mauss, Marcel, 40
Meddelelser om Grønland, 89, 98–9
Meinardus, Wilhelm, 21
Meteorological Institute,
  Denmark (*Meteorologiske Institut*), 117
meteorology, 3, 16, 19, 23, 24, 25, 26, 31, 33, 34–5, 36, 37, 45, 48, 49, 51, 56, 61, 63, 67, 91, 93, 97, 103, 105, 117, 118
metropolitan Denmark, 5, 13, 19, 78, 80, 98, 125n18
missiles, long range, 1, 2, 9, 100–2
Mock, Steven J., 106–7, 109, 110
Moltke, Count Carl, 118
Monroe Doctrine, 5
Munck, Ebbe, 55, 117, 118
Musée de l'Homme. *See* Musée du Trocadero
Musée du Trocadero, 39–40
Mylius-Erichsen, Ludvig, 23, 118

Nansen, Fridtjof, xvii, 3, 4, 8, 12, 15–16, 17
Narssarsuaq, xix, 20, 80
National Center for Scientific Research, France (*Centre National de la Recherche Scientifique*), 44, 88–9, 146n95
NATO (North Atlantic Treaty Organization), 5, 20, 62, 63, 79, 98, 101, 104, 117, 130–1n47, 158–9n7
Nestlé, 44
Nielsen, Niels, 97–9, 119, 178n26
Noe-Nygaard, Arne, 97–8
Nordenskiöld, Adolf Erik, xvii, 15, 128n23
Norse. *See* Greenland, history of:
  Norse settlements
North American Air Defense Command (NORAD), 20, 112
North American continental security/defense, 2, 9, 20, 61–2, 83, 112, 115
Northwest Passage, 18
Nørvang, Aksel, 80
Norway, 3, 5, 15, 17–18, 43, 65, 85, 102, 115–16, 118, 125n17, 127n15, 176n3
Norwegian Polar Institute (*Norsk Polarinstitutt*), 85
nunataks, 11, 14

Oeschger, Hans, 94–5, 114, 179n30
Operation Highjump, 109
Operation Icecap, 66
Operation King Dog, 66
Operation PCA 68, 81
Orain, Fred, 58
Østerbygd (Eastern Settlement), 14, 127n18

Paars, Claus Enevold, 8, 13–14, 18, 127n14
paleoclimatology. *See* climatology/climate
Partridge, Earle E., 109

INDEX

patronage/funding, 6, 7. *See also individual expeditions*
Peary, Robert E., xvii, 12, 13, 15, 18–19, 132n61
Peary's Channel, 132n61
Perez, Michel, 40
polar aviation. *See* aviation
polar cooling/warming, 53, 94, 96, 120
polar exploration. *See* Arctic: exploration
polar warming. *See* polar cooling/warming
polar whiteouts. *See* whiteouts
Pommier, Robert, 46, 69
Project Blue Ice, 174n132
Project Ice Skate, 104
Project Iceworm, 81
Project Jello
  conception and planning, 66, 70–1
  and Denmark, 78, 81–2
  and Eismitte, 1, 2, 9, 66–7, 70, 75–6, 83, 111–12
  and Expédition Glaciologique Internationale au Groënland (EGIG), 91
  and Expéditions Polaires Françaises (EPF), 70, 75–6
  food, 71–3, 76
  funding, 83
  ice cores, 76–8, 95
  living conditions, 71–3, 75
  logistics, 70–6
  military dimension, 67, 69, 83
  permission to work in Greenland, 81–2
  personnel. (*see individual names*)
  route, xix, 67
  scientific work, 67, 70, 76–8, 92, 106, 107, 154n74
  transportation technologies, 67, 70–1, 73–5, 173n121
Project Mint Julep, 66, 174n128
Project Overheat, 174n128
Project Snowman, 66, 109–10

Ragle, Richard H., 68, 72
Rasmussen, Gustav, 19, 79
rations, 33, 40, 49, 50–1, 71–3, 91, 103. *See also individual expeditions*
remote sensing, 9, 108, 112, 114
Renaud, André, 94
Rink, Hinrich, 12, 14–15
Roosevelt, Franklin D., 19, 104
Rosencrantz, Alfred, 97
Rouillon, Charles Gaston, 48, 56, 97
Royal Prussian Aeronautical Observatory (*Königlich-Preußische Aeronautische Observatorium*), 23, 26

Schleswig-Holstein, 158–9n7
Schmidt-Ott, Friedrich, 21–2, 132n57
Scholander, Per, 95
Scientific nationalism. *See* Denmark: scientific nationalism
sea level rise. *See* ice melt
Seeley, Stuart William, 101–2
seismology, 3, 21, 34, 36–7, 48, 49, 51, 53, 56, 63, 75, 87, 88, 90, 91, 92, 106
SHORAN, 101–2, 172n115. *See also* Aslakson, Carl I.; HIRAN; Seeley, Stuart William
Skinrood, Alan, 67, 73
Snow, Ice and Permafrost Research Establishment (SIPRE), 63–6, 68, 70, 71, 75, 76, 77, 82, 90, 91, 103–6, 179n30
snow markers, 52, 70, 73, 76, 92–3, 103, 106–7
snowfall. *See* Greenland, geography of: snowfall/accumulation
Søndre Strømfjord (now Kangerlussuaq Fjord), 19, 20, 62, 66, 79, 80, 91, 109
Sondrestrom Air Base, xix, 74, 82, 91, 104, 106, 113, 121
Sorge, Ernst, 26–38, 45, 47, 51–2, 53, 70, 76, 77, 120

Sorge, Ernst—*Continued*
Sorge's law of densification of snow, 36, 136n113
sovereignty, 7, 29, 85, 102
and Greenland/Denmark, 5, 18, 19, 23, 54–6, 78–82, 97, 115–17, 119–20, 130–1n47
intellectual/epistemological, 97, 116–17
Soviet Union, 20, 41, 43, 57, 61–2, 63, 65, 90, 112, 130–1n47, 146n93, 147n4, 173n122
Station Centrale. *See* Expéditions Polaires Françaises: Station Centrale
Station Eismitte. *See* Wegener expedition to Eismitte (1930–1931), The: Station Eismitte
Station Nord, 121
Stefánsson, Sigurdur, 18
Steinhauser, Peter, 91
Storgaard, Einar, 37
Svane, Aksel, 99
Svenningsen, Nils, 97–8
Sverdrup, Harald Ulrik, 85
Swedish-Danish-Norwegian expedition to Greenland, 54, 117–18
Swiss Commission for Snow and Avalanche Research, 85
Switzerland, 65, 85, 86, 87, 88, 89, 93, 94–5, 108, 113, 179–80n36

Taylor, Gérard, 42
Terre Adélie, 42–3, 45, 59, 90, 137–8n16
Base Dumont d'Urville, 90, 146n95
See also France: Antarctic claim
Three Year Expedition to East Greenland (1931–1934), 116
Thule Air Base, xix, 20, 66, 67, 68–9, 70–1, 74, 75, 76, 80, 81, 82, 99, 105, 110, 121, 175–6n2

Thule Culture. *See* Greenland, history of: Paleo-Eskimo cultures
Trudeau, Arthur G., 62
Tschaen, Louis, 107
Turner, William H., 109

Ulfsson, Gunnbjørn, 17
United States
Air Force, 2, 9, 20, 61, 62, 67, 69, 75, 82, 91, 97, 101–5, 106, 107, 109–10, 112–13, 121
Air Photographic and Charting Service, 103, 105
Arctic, Desert and Tropic Information Center (ADTIC), 41, 73
Army (*including* Army Corps of Engineers), 7, 47, 63, 64, 65, 67, 69, 72, 74, 75, 91, 100, 103, 105, 106, 111, 167–8n80
Army Air Forces, 41, 47, 101, 153n65
Army Arctic Construction and Frost Effects Laboratory, 63, 77
cold environments warfare, 63–4
Cold Regions Research and Engineering Laboratory (CRREL). *See* Cold Regions Research and Engineering Laboratory,
and Denmark, 19–20, 78–82, 97, 104, 117
Geological Survey, 34, 68, 86
and glaciology, 63–7, 76–7, 86, 89–90, 105, 113–14
and Greenland, attempted purchase of, 19
and Greenland, Cold War presence in, 2, 5, 9, 19–20, 61–7, 70–83
and Greenland, defense areas in, 20, 80–1, 97–8, 119
and Greenland, military facilities in, xix, 5, 9, 19–20, 54, 62, 66, 70, 80, 82, 97–8, 102,

112–14, 121, 150n24. (*see also individual air bases and military facilities*)
and Greenland, pre-World War II interest in, 18–19
and Greenland, weather stations in, 19, 35, 61, 117
and Greenland, World War II presence in, 5, 19, 35, 54, 78
military, 1, 2, 5, 9, 19, 35, 41–2, 54, 61–7, 69, 70, 73–4, 77, 78–83, 91, 96, 97, 101–3, 105, 109, 117, 146n93, 150n24. (*see also* United States: individual military units)
Natick Laboratories, 82
Navy, 15, 108, 109–10
polar strategy (*including* science-based), 2, 5, 61–6, 82–3
Quartermaster Food and Container Institute, 72
Quartermaster Subsistence Research and Development Laboratory, 71–2
Snow, Ice and Permafrost Research Establishment (SIPRE). (*see* Snow, Ice and Permafrost Research Establishment)
State Department, 75, 80–1
See also Greenland; Greenland, history of; *individual expeditions*; North American continental security
United States Virgin Islands. *See* Danish West Indies
University of Michigan expeditions to Greenland. *See* Greenland Expeditions of the University of Michigan (1926–1931)
Urey, Harold, 77
Uummannaq, 99

Vatnajökull ice dome, 59, 85
Vaugelade, Jean, 120
Victor, Paul-Emile, 39–40, 107
and anthropology, 39–41
and Denmark, 54–6, 96–100, 118
and Eismitte, 2, 13, 39
and Expédition Glaciologique Internationale au Groënland, 87–9, 91, 96–100, 158–9n7, 159n10
and Expéditions Polaires Françaises, 8–9, 42–5, 47, 49, 51, 54–9, 87, 137n12
interwar expeditions to Greenland, 39–41
and Polynesia, 136n2
promotion/publicity, 8–9, 40, 41, 44, 57–9, 100
technological vision, 40, 42, 47, 57
and the United States, 41–2, 44, 47, 58, 70, 75–6, 91
during World War II, 41–2
Villumsen, Rasmus, 2, 28–30, 175–6n2

Wallerstein, George, 67, 81, 107, 154n74
Washburn, Albert L., 91
Watkins, Gino, 133n69
Wærum, Ejnar, 97–8
weasels (polar tractors), 8, 42, 46–7, 52, 56, 57, 67, 70–1, 74, 83, 87, 90, 91, 108
weather modification, 105, 170n102
Webb, Byron B., 91
Wegener, Alfred, 22–24
and continental drift, 22, 24
Danmark expedition (1906–1908), 23
death, 2, 20, 26, 28–30, 38, 108
Eismitte expedition (1930–1931), 1, 3, 8, 20, 22, 24–30, 36, 108
Greenland expedition (1912–1913), 16–17, 24
Wegener expedition to Eismitte (1930–1931), The
conception and planning, 1, 3, 22, 24–7
and Denmark, 25

Wegener expedition—*Continued*
  and Eismitte, 1, 8, 20, 25, 27,
    37–8, 53, 108, 111–12
  food, 33
  funding, 6, 22
  legacy, 2, 34–7, 45–6, 51, 70, 76,
    120
  overwinter living conditions,
    29–33
  personnel. (*see individual names*)
  scientific work, 8, 22, 25, 31,
    33–7, 49, 53, 106, 135n106
  Station Eismitte, 25, 26–34,
    36–7, 45, 47, 48, 175n146
  technological vision, 29–30
  transportation technologies
    (*including* propeller sledges),
    21–2, 25, 27, 29–30
  Wegener's final journey, 27–9

Wegener, Kurt, 23, 29, 37
Weiken, Karl, 28, 29
Weimar Republic, 6, 21–2, 115
whiteouts, 105, 110. *See also* weather
  modification
Wilson, Charles R., 64
World Geodetic System 1960 (WGS
  60), 102
World War I, 7, 17, 21, 24, 36, 115
World War II, 2, 39, 40, 41–2, 47,
  52, 63, 64, 68, 71, 75, 87,
  98–9, 109, 115, 117, 138n19
  Denmark, German occupation of,
    5, 19, 78, 97, 98, 155n79
  and geodesy, 100, 101
  Greenland, strategic significance
    of, 5, 19, 34–5
  Greenland, United States presence
    in, 5, 19, 35, 54, 78

The manufacturer's authorised representative in the EU is Springer Nature Customer Service Centre GmbH, Europaplatz 3, 69115 Heidelberg, Germany. If you have any concerns regarding our products, please contact ProductSafety@springernature.com

Printed and bound by CPI Group (UK) Ltd, Croydon, CR0 4YY

23/03/2026

02076449-0009